Mathematical Aspects in
Non-Equilibrium Thermodynamics

Mathematical Aspects in Non-Equilibrium Thermodynamics

Editors

Róbert Kovács
Patrizia Rogolino
Francesco Oliveri

MDPI • Basel • Beijing • Wuhan • Barcelona • Belgrade • Manchester • Tokyo • Cluj • Tianjin

Editors

Róbert Kovács
Wigner Research Institute for
Physics
Hungary

Patrizia Rogolino
Università di Messina
Italy

Francesco Oliveri
Università di Messina
Italy

Editorial Office
MDPI
St. Alban-Anlage 66
4052 Basel, Switzerland

This is a reprint of articles from the Special Issue published online in the open access journal *Symmetry* (ISSN 2073-8994) (available at: https://www.mdpi.com/journal/symmetry/special_issues/Mathematical_Aspects_Non-equilibrium_Thermodynamics).

For citation purposes, cite each article independently as indicated on the article page online and as indicated below:

LastName, A.A.; LastName, B.B.; LastName, C.C. Article Title. *Journal Name* **Year**, *Volume Number*, Page Range.

ISBN 978-3-0365-7762-3 (Hbk)
ISBN 978-3-0365-7763-0 (PDF)

Contents

About the Editors

Róbert Kovács

Róbert Kovács has a degree in mechanical engineering, graduated in 2015, and then obtained the Ph.D in 2017 in Budapest University of Technology and Economics (BME), in collaboration with the Hungarian Academy of Sciences. He is currently working on continuum thermodynamics, both on experimental and theoretical problems, including the development of numerical schemes, analytical solutions of evolution equations, and other mathematical aspects. He has been a visting researcher at Northeastern University, MA, USA; at Messina University, Italy; and at Université du Québec à Chicoutimi, Québec, Canada.

Patrizia Rogolino

Patrizia Rogolino is Professor of mathematical physics at the University of Messina. Her research interests include irreversible thermodynamics of continuous media, nonlinear theories of heat conduction, and thermoelectric phenomena in nanosystems. She is the author of several scientific papers on peer reviewed journals.

Francesco Oliveri

Francesco Oliveri is full professor of mathematical physics at the University of Messina. His research interests include mathematical apparatus of quantum mechanics for the modelization of macroscopic systems, Lie symmetries of differential equations, thermodynamics of continuous media with nonlocal constitutive relations, computer algebra. He is the author of numerous scientific papers on these topics in peer-reviewed journals as well author and editor of some volumes.

Editorial

Mathematical Aspects in Non-Equilibrium Thermodynamics

Róbert Kovács [1,2,3,*], Patrizia Rogolino [4] and Francesco Oliveri [4]

1 Department of Energy Engineering, Faculty of Mechanical Engineering, Budapest University of Technology and Economics, 1111 Budapest, Hungary
2 Department of Theoretical Physics, Wigner Research Center of Physics, Konkoly-Thege M. 29-33, 1121 Budapest, Hungary
3 Montavid Thermodynamic Research Group, 1112 Budapest, Hungary
4 Department of Mathematical and Computer Sciences, Physical Sciences and Earth Sciences, University of Messina, Viale F. Stagno d'Alcontres 31, 98166 Messina, Italy
* Correspondence: kovacs.robert@wigner.hu

Introduction

Prof. Csaba Asszonyi, D.Sc. (1941–2022): The present Special Issue is dedicated to the memory of our beloved, respected friend, colleague and teacher, the late Professor Csaba Asszonyi.

The research field of professor Asszonyi was continuum mechanics and irreversible thermodynamics. He played a pioneering role in establishing the thermodynamical background of rock rheology and engineering rock mechanics. He was educated as a mechanical engineer, started his career as research engineer, and then performed coordinated mining research in Hungary. Later on, he went into the industry and became a company group leader. Seventeen years ago, he established the Montavid Thermodynamic Research Group.

His thinking focused on thermodynamic concepts, connecting an application-oriented, engineering attitude with deep theoretical ideas. He developed several industrial applications of thermodynamic rheology. His contributions included the extension of linear viscoelasticity with internal variables and the unification of classical rheological bodies in a thermodynamic framework. He was the author of more than two hundred articles, dozens of patents, and ten books. He refused honours and distinctions, and only at the end of his life became the honorary president of the Society for the Unity of Science and Technology.

The Special Issue "Mathematical Aspects in Non-equilibrium Thermodynamics" consists of five original research papers. Although the current topic has a long history, there are still numerous open questions regarding the structure of evolution equations, the corresponding thermodynamically compatible initial and boundary conditions, and also their relation to experimental and practical aspects. These five papers actually cover various recent and relevant topics such as optimization, finite time thermodynamics, the role of the second law in continuum physics, multi-component mixtures, and boundary conditions. We hope that this Special Issue will be able to play a role in further progress to come in the future.

In the paper "The Role of the Second Law of Thermodynamics in Continuum Physics: A Muschik and Ehrentraut Theorem Revisited" by V. A. Cimmelli and P. Rogolino [1], the authors revisited the second law of thermodynamics and how the entropy inequality plays a crucial role in the derivation of evolution equations, also providing local and global formulations of the second law. The classical results of Muschik and Ehrentraut are reformulated in the present modern mathematical context of second law, thus highlighting a few geometric aspects as an outcome. They also emphasized that the non-equilibrium concept of temperature and entropy far from equilibrium is not necessarily identical to the one close to equilibrium, and how these notions need further investigation.

The paper "Integrability of the Multi-Species TASEP with Species-Dependent Rates", written by Eunghyun Lee [2], is related to totally asymmetrical simple exclusion processes; it

Citation: Kovács, R.; Rogolino, P.; Oliveri, F. Mathematical Aspects in Non-Equilibrium Thermodynamics. *Symmetry* 2023, 15, 929. https://doi.org/10.3390/sym15040929

Received: 2 March 2023
Accepted: 10 April 2023
Published: 17 April 2023

was shown that the Bethe ansatz method is applicable to processes in the present N-particle mixtures with species-dependent rates, providing transition probabilities for all possible compositions of species. Despite the limitations detailed in [2], this study serves as a basis for future investigation to see if the methods can be used to study the species inhomogeneity of other multi-species models.

In the paper "Shock Structure and Relaxation in the Multi-Component Mixture of Euler Fluids" by Madjarevic et al. [3], an important benchmark study related to shock structures is presented. Here, the authors utilized a multi-component mixture of Euler fluids, whose evolution equations possess a hyperbolic structure, originating from extended thermodynamics. The present study is concerned with a three-component mixture of polyatomic gases inheriting the kinetic theory formulation for the phenomenological coefficients. The the quantitative characteristics of the shock profiles, such as the temperature overshoot, the shock thickness, and the resulting relaxation times were investigated, thus providing a deeper insight into a complex, coupled phenomenon.

The paper "Cyclic Control Optimization Algorithm for Stirling Engines" by Raphael Paul and Karl Heinz Hoffmann [4] deals with an optimization problem related to non-equilibrium Stirling engines. The authors focused their attention on the optimization of both the power and efficiency, using an indirect iterative gradient algorithm. The problem formulation led to a particular Hamiltonian system, describing attractive and repulsive limit cycles, with periodic boundary conditions. They provided detailed insight into the problem formulation and optimization algorithm, and therefore their results are of high importance in dealing with similar optimization tasks for other thermodynamic cycles.

The last paper of the present Special Issue, titled "Recent Advances on Boundary Conditions for Equations in Nonequilibrium Thermodynamics" [5], is written by Wen-An Yong and Yizhou Zhou. They focused on linearized systems obeying the hyperbolic structure originating from extended thermodynamics and reviewed the possible (proper) boundary conditions in the light of uniform and generalized Kreiss conditions. The structural stability of the studied PDEs was also satisfied. As these conditions are strongly related to the suitability of hyperbolic equations, the present results could serve as a future basis for the comparison of various thermodynamic approaches and provide hints to extend the present formalism to nonlinear problems.

Acknowledgments: We would like to express our gratitude to the Editorial Board of Symmetry for they helpful attitude and also to the Authors who made this Special Issue successful.

Conflicts of Interest: The authors declare no conflict of interest.

References

1. Cimmelli, V.A.; Rogolino, P. The Role of the Second Law of Thermodynamics in Continuum Physics: A Muschik and Ehrentraut Theorem Revisited. *Symmetry* **2022**, *14*, 763. [CrossRef]
2. Lee, E. Integrability of the Multi-Species TASEP with Species-Dependent Rates. *Symmetry* **2021**, *13*, 1578. [CrossRef]
3. Madjarević, D.; Pavić-Colić, M.; Simić, S. Shock Structure and Relaxation in the Multi-Component Mixture of Euler Fluids. *Symmetry* **2021**, *13*, 955. [CrossRef]
4. Paul, R.; Hoffmann, K.H. Cyclic Control Optimization Algorithm for Stirling Engines. *Symmetry* **2021**, *13*, 873. [CrossRef]
5. Yong, W.-A.; Zhou, Y. Recent Advances on Boundary Conditions for Equations in Nonequilibrium Thermodynamics. *Symmetry* **2021**, *13*, 1710. [CrossRef]

Article

The Role of the Second Law of Thermodynamics in Continuum Physics: A Muschik and Ehrentraut Theorem Revisited

Vito Antonio Cimmelli [1,*,†] and Patrizia Rogolino [2,†]

1   Department of Mathematics, Computer Science and Economics, University of Basilicata, Viale dell'Ateneo Lucano n. 10, 85100 Potenza, Italy
2   Department of Mathematical and Computer Sciences, Physical Sciences and Earth Sciences, University of Messina, Viale F. Stagno d'Alcontres, 31, 98166 Messina, Italy; progolino@unime.it
*   Correspondence: vito.cimmelli@unibas.it
†   These authors contributed equally to this work.

Abstract: In continuum physics, constitutive equations model the material properties of physical systems. In those equations, material symmetry is taken into account by applying suitable representation theorems for symmetric and/or isotropic functions. Such mathematical representations must be in accordance with the second law of thermodynamics, which imposes that, in any thermodynamic process, the entropy production must be nonnegative. This requirement is fulfilled by assigning the constitutive equations in a form that guaranties that second law of thermodynamics is satisfied along arbitrary processes. Such an approach, in practice regards the second law of thermodynamics as a restriction on the constitutive equations, which must guarantee that any solution of the balance laws also satisfy the entropy inequality. This is a useful operative assumption, but not a consequence of general physical laws. Indeed, a different point of view, which regards the second law of thermodynamics as a restriction on the thermodynamic processes, i.e., on the solutions of the system of balance laws, is possible. This is tantamount to assuming that there are solutions of the balance laws that satisfy the entropy inequality, and solutions that do not satisfy it. In order to decide what is the correct approach, Muschik and Ehrentraut in 1996, postulated an amendment to the second law, which makes explicit the evident (but rather hidden) assumption that, in any point of the body, the entropy production is zero if, and only if, this point is a thermodynamic equilibrium. Then they proved that, given the amendment, the second law of thermodynamics is necessarily a restriction on the constitutive equations and not on the thermodynamic processes. In the present paper, we revisit their proof, lighting up some geometric aspects that were hidden in therein. Moreover, we propose an alternative formulation of the second law of thermodynamics, which incorporates the amendment. In this way we make this important result more intuitive and easily accessible to a wider audience.

Keywords: second law of thermodynamics; dissipation principle; state space; balance laws; entropy inequality

Citation: Cimmelli, V.A.; Rogolino, P. The Role of the Second Law of Thermodynamics in Continuum Physics: A Muschik and Ehrentraut Theorem Revisited. *Symmetry* **2022**, *14*, 763. https://doi.org/10.3390/sym14040763

Academic Editors: Alessandro Sarracino and Kazuharu Bamba

Received: 2 February 2022
Accepted: 1 April 2022
Published: 7 April 2022

Publisher's Note: MDPI stays neutral with regard to jurisdictional claims in published maps and institutional affiliations.

1. Introduction

Let B be a continuous body undergoing a thermomechanical transformation, whose evolution in the spacetime is ruled by the system of balance laws

$$U_{\beta,t} + U_{\beta,j} v_j + \Phi^{\beta}_{k,k} = r_\beta, \quad \beta = 1 \ldots \omega, \tag{1}$$

with v_j as the components of the velocity field on B entering the total time derivative, Φ^{β}_k as the components of the flux of U_β, and r_β as the production of U_β (for the sake of simplicity, we assume that the supplies are zero). Moreover, the symbols $f_{,t}$ and $f_{,j}$ mean the partial derivatives of function f with respect to time and to the spatial coordinate x_j, $j = 1, 2, 3$, respectively.

For instance, in classical rational thermodynamics, Equation (1) is the balance of mass, linear momentum, angular momentum, and energy [1,2],

$$\rho_{,t} + \rho_{,i}v_i + \rho v_{j,j} = 0, \tag{2}$$

$$\rho v_{i,t} + \rho v_{i,j}v_j - T_{ij,j} = \rho b_i, \tag{3}$$

$$T_{ij} = T_{ji}, \tag{4}$$

$$\rho\varepsilon_{,t} + \rho\varepsilon_{,j}v_j - T_{ij}v_{i,j} + q_{j,j} = \rho r, \tag{5}$$

where ρ denotes the mass density, ε the specific internal energy, r the energy supply, v_i, q_i, and b_i the components of velocity, heat flux, and body force, respectively, and T_{ij} are the components of the Cauchy stress [1,2], while in the extended non-equilibrium thermodynamic theories, taking the fluxes as independent variables, the set of field equations includes the balance laws for the independent fluxes [3–7].

We suppose that the fields U_β, the fluxes Φ_k^β, and the productions r_β depend on ω unknown fields $z_\alpha(x_j, t)$ and on their spatial derivatives $z_{\alpha,j}(x_j, t)$. Then, suitable constitutive equations must be assigned for them. Such equations must account for material symmetry, namely, the invariance of certain physical properties with respect to a group of coordinate transformations (symmetry group). Thus, suitable representation theorems for symmetric and/or isotropic functions must be applied [8]. Through the constitutive equations, material symmetry reflects itself in the Equation (1) field, which inherit from them particular mathematical properties. Meaningful examples of such reciprocal influences are illustrated in [9], where some important equations of continuum physics are studied by the Lie symmetry analysis of differential equations. On the other hand, the mathematical representation of material symmetry cannot neglect the constraint dictated by the second law of thermodynamics, which imposes the local entropy production

$$\sigma^{(s)} = \rho s_{,t} + \rho s_{,j}v_j + J_{k,k} - \rho(r/\vartheta), \tag{6}$$

where s is the specific entropy, J_k are the components of the entropy flux, and ϑ the absolute temperature, are nonnegative, namely,

$$\sigma^{(s)} \geq 0, \tag{7}$$

whatever the thermodynamic process is [1,2]. Indeed, the unilateral differential constraint (7) can be interpreted either as a restriction on the solutions of Equation (1), or as a restriction on the constitutive equations that characterize the fields U_β, Φ_k^β, r_β, s, and J_k. In the first case, it leads to the following assumption:

Assumption 1. *Among the mathematical solutions of Equation (1), we must find those that are physically realizable by substituting them into the Inequality (7) and checking the sign.*

In the second case, instead, it leads to a different assumption, namely:

Assumption 2. *The constitutive equations for U_β, Φ_k^β, r_β, s, and J_k must be assigned in such a way that the constraint (7) is satisfied along arbitrary thermodynamic processes.*

Modern constitutive theories of continuum thermodynamics are based on the second statement, which was formulated for the first time by Coleman and Noll [1] in 1963, and is universally known as the entropy (or dissipation) principle [10]. On the other hand, since the determination of solutions of Equation (1) is, in general, very difficult, the Coleman–Noll approach is the most convenient one used for determining the consequences of the second law. Two rigorous mathematical procedures for analyzing the restrictions dictated by the second law on the constitutive equations are based on such an assumption, namely, the Coleman–Noll and Liu procedures [1,11,12].

In order to investigate if the entropy principle, as formulated by Coleman and Noll, is just an operative assumption or a consequence of a general physical law, Muschik and Ehrentraut in 1996 proved their celebrated theorem in [13], which we revisit here within a geometric perspective. Since the theorem has a complex mathematical formulation, herein we limit ourselves to provide a sketch of the result, referring the reader to reference [13] for details. Their starting point is the following concept of equilibrium for a thermodynamic system.

> *A thermodynamic system is said to be in equilibrium (stable or metastable), if isolation of the system from its environment does not affect the observables of the system (or in other words, if the state of the isolated system is the same as the state of the system prior to insulation).*

Then, they define the process direction vectors as those vectors that are tangent to the curve representative of a process in the state space. Moreover, they show that in a fixed point of such a curve, the entropy production is linear in the process direction vectors y_α. The vectors y_α are said irreversible, if $\sigma^{(s)}(y_\alpha) > 0$, reversible, if $\sigma^{(s)}(y_\alpha) = 0$, non-real, if $\sigma^{(s)}(y_\alpha) < 0$. In particular, a vector y_α such that $\sigma^{(s)}(y_\alpha) = 0$, is called the reversible process direction.

At this point, Muschik and Ehrentraut prove their fundamental.

Proposition 1. *If in a point of non-equilibrium of the curve representative of the process in the state space there exist both irreversible and non-real process directions, then, in that point, it is possible to obtain a reversible process direction as a linear combination of the latter.*

Such a result is unphysical, since in a non-equilibrium point the entropy production can be zero. In order to overcome such a discrepancy, Muschik and Ehrentraut add to the classical formulation of the second law, represented by the Inequality (7), the following amendment.

> *Except in equilibria, reversible process directions in state space do not exist.*

The consequences of the amendment are severe, because it implies that, in a given point, the process directions are either all irreversible (or reversible, in the limit of quasi-static transformation), or all non-real. On the other hand, since non-real process directions do not exist in nature, we must choose the first option. As a consequence, non-real process directions that must be excluded by the second law do not exist. Moreover, since the point on the curve of the process is arbitrary, we are led to conclude that there are no non-real thermodynamic processes to be forbidden by the second law and, hence, the last necessarily restricts the constitutive equations and not the thermodynamic processes. In this way, the classical Coleman–Noll approach follows by a rigorous proof.

It is worth observing that Proposition 1 and its consequences are not aimed at giving more deep physical insight on some basic concepts of thermodynamics, such as equilibrium and reversibility, since in [13] those concepts are the classical ones. Indeed, Proposition 1 aims at answering the following question, which is fundamental from the methodological point of view: is the Coleman–Noll interpretation of the second law, as a constraint on the constitutive equations, a mathematical consequence of the basic laws of thermodynamics or an additional operative assumption that does not follow from those laws? The answer, given by Proposition 1, is that such an interpretation is a consequence of the second law, provided the amendment is postulated. In the absence of the amendment, the Coleman–Noll approach would be an arbitrary assumption and, hence, it could be questioned. The present paper is motivated by the observation that the important result illustrated above can be put in a more general and accessible form within a geometric framework. To achieve that task, we use the results in references [14,15], where a geometric perspective on non-equilibrium thermodynamics has been given. The chosen state space is different with respect to that considered reference [13], because we do not include in it the time derivatives. In this way, the constitutive equations we deal with are suitable to satisfy the principle of material indifference [2]. After defining the space of the higher derivatives, we introduce the definitions of the real, ideal, and over-ideal vector of the higher derivatives. For thermodynamic processes, we give the definitions of irreversible, reversible, and over-reversible process, by analyzing the properties of its representative curve in the fiber bundle of the configuration spaces. Once the geometric framework is complete, we reformulate the second law of thermodynamics, both locally and globally, in space and time, in order to encompass the amendment. In this way, we are able to prove a new formulation of the Muschik and Ehrentraut theorem. In the discussion in Section 5, we highlight some of the main advantages of the geometric approach presented here. We underline that our generalized formulation of the second law seems to be more adapt for some recent thermodynamic theories that analyze real transformations, which occur in a finite time, and not by quasi-static transformations, which require an infinite time to be realized [16–18]. Moreover, we will see that the use of the fiber bundle allows to provide, in a very intuitive manner, the mathematical definition of reversible and irreversible processes, and to generalize, in a natural way, local results to global ones. Finally, our approach leads to a physically sound interpretation of the principle of local equilibrium [19], which provides the theoretical justification of the definition of entropy and temperature outside the equilibrium.

The paper runs as follows.

In Section 2, we construct a new thermodynamic framework for non-equilibrium processes. In Section 3, we present a new formulation, both locally and globally, of the second law of thermodynamics. In Section 4, we prove the Muschik and Ehrentraut theorem. In Section 5, we discuss our results and present some open problems that will be considered in future research.

2. The Thermodynamic Framework

In this section, we construct a geometric framework where our main results can be formulated. To this end, we begin by providing some basic definitions.

Definition 1. *The space C_t of the configurations at the instant t is represented by a ω-dimensional vector space spanned by the solutions $z_\alpha(x_j, t)$ of Equation (1) with a structure of a finite-dimensional manifold.*

We assume that the total configuration space is given by the disjointed union

$$\mathcal{C} = \bigcup_{t \in [0, \infty]} \{t\} \times C_t, \tag{8}$$

with a given natural structure of a fiber bundle over the real line $\mathbb{R}$ where time flows [14,15].

Definition 2. *$\mathcal{C}$ is called the configuration bundle.*

Under the natural assumption that C_t does not vary in time, namely, $C_t = C \,\forall t$, then $\mathcal{C}$ has the topology of the Cartesian product

$$\mathcal{C} = \mathbb{R} \times C. \tag{9}$$

Definition 3. *A vector valued function $\pi : t \in [\tau_0, \tau_0 + \tau] \subseteq \mathbb{R} \to z_\alpha(x_j, t) \in \mathcal{C}$ is said to be a thermodynamic process π of duration τ. Moreover, $\pi = \pi(t)$ is the parametric equation of the curve Γ representative of π in $\mathcal{C}$.*

Definition 4. *For $t_0 \in [\tau_0, \tau_0 + \tau]$, a vector valued function $p : t \in [t_0, \tau_0 + \tau] \subseteq \mathbb{R} \to z_\alpha(x_j, t) \in \mathcal{C}$ is said to be a restricted thermodynamic process p of initial point t_0 and duration $\tau_0 + \tau - t_0$, reference [14]. Moreover, $p = p(t)$ is the parametric equation of the curve γ representative of p in $\mathcal{C}$.*

Remark 1. *For $t_0 = \tau_0$ we have $p(t) = \pi(t)$, for $t_0 = \tau_0 + \tau$, $p(t)$ is the process of duration 0, i.e., the null process.*

As said in Section 2, in order to find the fields $z_\alpha(x_j, t)$, i.e., to solve the system (1), for the quantities U_β, Φ_k^α and r_α constitutive equations must be assigned on a suitable state space.

Definition 5. *The 4ω-dimensional vector space with the structure of a finite-dimensional manifold*

$$\Sigma_t = \left\{ z_\alpha(x_j, t),\, z_{\alpha,j}(x_j, t) \right\}. \tag{10}$$

for any value of the time variable t, it represents a local in the time state space and it is called state space at the instant t.

Definition 6. *The disjoint union*

$$\mathcal{S} = \bigcup_{t \in [0, \infty]} \{t\} \times \Sigma_t, \tag{11}$$

with a given natural structure of a fiber bundle over the real line $\mathbb{R}$ where time flows, it represents the total configuration space and it is said to be the thermodynamic bundle.

Again, under the natural assumption that Σ_t does not vary in time, namely, $\Sigma_t = \Sigma \,\forall t$, then $\mathcal{S}$ has the topology of the Cartesian product

$$\mathcal{S} = \mathbb{R} \times \Sigma. \tag{12}$$

Of course,

$$C_t \subset \Sigma_t, \quad \mathcal{C} \subset \mathcal{S}. \tag{13}$$

The balance Equation (1) on the local in time state space Σ_t reads

$$\frac{\partial U_\beta}{\partial z_\alpha} z_{\alpha,t} + \frac{\partial U_\beta}{\partial z_{\alpha,j}} z_{\alpha,jt} + \frac{\partial U_\beta}{\partial z_\alpha} z_{\alpha,j} v_j + \frac{\partial U_\beta}{\partial z_{\alpha,k}} z_{\alpha,kj} v_j + \frac{\partial \Phi_k^\beta}{\partial z_\alpha} z_{\alpha,k} + \frac{\partial \Phi_k^\beta}{\partial z_{\alpha,j}} z_{\alpha,jk} = r_\beta. \tag{14}$$

In Equation (14), we may individuate the 10ω higher derivatives $\left\{ z_{\alpha,t},\, z_{\alpha,jt},\, z_{\alpha,jk} \right\}$, which are the space and time derivatives of the elements of Σ_t.

Definition 7. *The local in time 10ω-dimensional vector space*

$$H_t = \left\{ z_{\alpha,t}(x_j, t),\, z_{\alpha,jt}(x_j, t),\, z_{\alpha,jk}(x_j, t) \right\}, \tag{15}$$

and the fiber bundle

$$\mathcal{H} = \bigcup_{t \in [0, \infty]} \{t\} \times H_t. \tag{16}$$

represent the space of the higher derivatives at time t and its fiber bundle, respectively.

Definition 8. *A point $P \in B$ is said to be in a state of equilibrium at the instant t, if*

$$z_{\alpha,t}(P, t) + z_{\alpha,j}(P, t) v_j = 0 \ \forall\, t,$$

and

$$z_{\alpha,jt}(P, t) + z_{\alpha,jk}(P, t) v_k = 0 \ \forall\, t.$$

The Definition 8 leads to

Definition 9. *The equilibrium subspace of H_t and its fiber bundle are given by*

$$\hat{E}_t = \left\{ z_{\alpha,jk}(x_j,\, t) \right\}, \tag{17}$$

and

$$\hat{\mathcal{E}} = \bigcup_{t \in [0,\infty]} \{t\} \times \hat{E}_t. \tag{18}$$

Analogously, the entropy inequality on the state space reads

$$\rho \frac{\partial s}{\partial z_\alpha} z_{\alpha,t} + \rho \frac{\partial s}{\partial z_{\alpha,j}} z_{\alpha,jt} + \rho \frac{\partial s}{\partial z_\alpha} z_{\alpha,j} v_j + \rho \frac{\partial s}{\partial z_{\alpha,k}} z_{\alpha,kj} v_j + \frac{\partial J_k}{\partial z_\alpha} z_{\alpha,k} + \frac{\partial J_k}{\partial z_{\alpha,j}} z_{\alpha,jk} \geq 0. \tag{19}$$

Definition 10. *The local in time 10ω-dimensional vector space at time t*

$$W_t = \left\{ z_{\alpha,t}(x_j,\, t),\, z_{\alpha,jt}(x_j,\, t),\, z_{\alpha,jk}(x_j,\, t) \right\}, \tag{20}$$

and the fiber bundle

$$\mathcal{W} = \bigcup_{t \in [0,\infty]} \{t\} \times W_t. \tag{21}$$

define the vector space and the fiber bundle of the higher derivatives, respectively, whose state vectors satisfy the entropy inequality. Moreover, the equilibrium subspace of W_t and its fiber bundle are given by

$$E_t = \left\{ z_{\alpha,jk}(x_j,\, t) \right\}, \tag{22}$$

and

$$\mathcal{E} = \bigcup_{t \in [0,\infty]} \{t\} \times E_t. \tag{23}$$

Remark 2. *We defined two different spaces of the higher derivatives, one for the balance equations and another for the entropy inequality, related to the fundamental focus of the present investigation, namely, to determine the conditions, if any, under which all of the solutions of the balance laws are also solutions of the entropy inequality. This will be discussed in detail in the next section.*

The relations in Equations (14) and (19) can be arranged as follows

$$\frac{\partial U_\beta}{\partial z_\alpha} z_{\alpha,t} + \frac{\partial U_\beta}{\partial z_{\alpha,j}} z_{\alpha,jt} + \left(\frac{\partial U_\beta}{\partial z_{\alpha,k}} v_j + \frac{\partial \Phi_j^\beta}{\partial z_{\alpha,k}} \right) z_{\alpha,kj} = r_\beta - \frac{\partial U_\beta}{\partial z_\alpha} z_{\alpha,j} v_j - \frac{\partial \Phi_j^\beta}{\partial z_\alpha} z_{\alpha,j}. \tag{24}$$

$$\rho \frac{\partial s}{\partial z_\alpha} z_{\alpha,t} + \rho \frac{\partial s}{\partial z_{\alpha,j}} z_{\alpha,jt} + \left(\rho \frac{\partial s}{\partial z_{\alpha,k}} v_i + \frac{\partial J_i}{\partial z_{\alpha,k}} \right) z_{\alpha,ki} \geq -\rho \frac{\partial s}{\partial z_\alpha} z_{\alpha,j} v_j - \frac{\partial J_i}{\partial z_\alpha} z_{\alpha,i}. \tag{25}$$

Let us now define the $10\omega \times 1$ column vector function

$$y_\alpha \equiv \left(z_{\alpha,t},\, z_{\alpha,jt},\, z_{\alpha,kj} \right)^T, \tag{26}$$

the $\omega \times 1$ column vector

$$C_\beta \equiv r_\beta - \frac{\partial U_\beta}{\partial z_\alpha} z_{\alpha,j} v_j - \frac{\partial \Phi_j^\beta}{\partial z_\alpha} z_{\alpha,j},\quad \beta = 1 \ldots \omega, \tag{27}$$

and the $\omega \times 10\omega$ matrix

$$A_{\beta\alpha} \equiv \left[\frac{\partial U_\beta}{\partial z_\alpha}, \; \frac{\partial U_\beta}{\partial z_{\alpha,j}}, \; \left(\frac{\partial U_\beta}{\partial z_{\alpha,k}} v_j + \frac{\partial \Phi_j^\beta}{\partial z_{\alpha,k}} \right) \right], \quad j,k = 1,2,3, \tag{28}$$

with C_β and $A_{\beta\alpha}$ defined on $\mathcal{S}$. In this way, the balance Equation (24) can be rearranged as

$$A_{\beta\alpha}(\mathcal{S})y_\alpha = C_\beta(\mathcal{S}). \tag{29}$$

Analogously, after defining the $10\omega \times 1$ column vector function

$$B_\alpha(\mathcal{S}) \equiv \left(\rho \frac{\partial s}{\partial z_\alpha}, \; \rho \frac{\partial s}{\partial z_{\alpha,j}}, \; \left(\rho \frac{\partial s}{\partial z_{\alpha,k}} v_i + \frac{\partial J_i}{\partial z_{\alpha,k}} \right) \right)^T, \tag{30}$$

and the scalar function

$$D(\mathcal{S}) \equiv -\rho \frac{\partial s}{\partial z_\alpha} z_{\alpha,j} v_j - \frac{\partial J_i}{\partial z_\alpha} z_{\alpha,i}, \tag{31}$$

we can write the Inequality (25) as

$$B_\alpha(\mathcal{S})y_\alpha \geq D(\mathcal{S}). \tag{32}$$

Remark 3. *It is worth observing that the higher derivatives entering the system* (29) *are elements of H_t, while those entering the Inequality* (32) *are elements of W_t.*

From now on, we pursue our analysis under the hypothesis that B occupies the whole space. Then, for arbitrary $t_0 \in [\tau_0, \tau_0 + \tau]$, we consider the restricted process p of the initial instant t_0 and duration $\tau_0 + \tau - t_0$, and suppose that it corresponds to the solution of the Cauchy problem for the system (29) with initial conditions

$$z_\alpha(x_j, t_0) = z_{\alpha 0}(x_j), \quad \forall P \in C. \tag{33}$$

If $A_{\beta\alpha}$ and C_β are regular, and $A_{\beta\alpha}$ is invertible, the theorem of Cauchy–Kovalevskaya ensures that the Cauchy problem (29) and (33) has a unique solution continuously depending on the initial data (33), reference [20]. However, such a solution does not necessarily correspond to a thermodynamic process that is physically realizable, since the physically admissible solutions of (29) and (33) are only those solutions, which additionally satisfy the unilateral differential constraint (32). On the other hand, the problems (29) and (33) are very difficult to solve, in general, so that finding a solution, and verifying ex post if it also satisfies (32), does not seem to be a convenient procedure. For that reason, Coleman and Noll [1] in 1963 postulated the constitutive principle referred in Section 2, reference [10]. It is important to investigate if the Coleman and Noll postulate is a consequence of a general physical law or an arbitrary, although very useful, assumption, as observed by Muschik and Ehrentraut [13]. Such a study will be carried out in the following sections.

3. Local and Global Formulations of the Second Law of Thermodynamics

Let us consider now a fixed point $P_0 \in B$ whose vector position will be indicated by x_0, a fixed instant of time $t_0 \in [\tau_0, \tau_0 + \tau]$. We note that, whatever is t_0, it can can ever be considered as the initial time of a restricted process of duration $\tau_0 + \tau - t_0$. Moreover, let Σ_0, H_0, and E_0 the vector spaces $\Sigma_t(P_0, t_0)$, $H_t(P_0, t_0)$, and $E_t(P_0, t_0)$. When evaluated in (P_0, t_0), the balance Equation (29) and the entropy Inequality (32) transform in the algebraic relations:

$$A_{\beta\alpha}(\Sigma_0)y_\alpha = C_\beta(\Sigma_0), \tag{34}$$

$$B_\alpha(\Sigma_0)y_\alpha \geq D(\Sigma_0). \tag{35}$$

In this way, we can regard the $\omega \times 10\omega$ matrix $A_{\beta\alpha}(\Sigma_0)$ as a linear morphism from H_0 to the ω-dimensional Euclidean vector space defined on Σ_0. Analogously, the vector $B_\alpha(\Sigma_0)$ can be regarded as a linear application from H_0 in $\mathbb{R}$, so that $B_\alpha(\Sigma_0)$ belongs to the dual space H_0^* of H_0. It is worth observing that, since $A_{\beta\alpha}$ is supposed to be invertible (otherwise the Cauchy problem (29) and (33) would not admit a unique solution), the algebraic relations (34) allow to determine ω of the 10ω components of y_α. Moreover, by spatial derivation of the initial conditions (33), we have

$$z_{\alpha,jk}(x_j, t_0) = z_{\alpha 0,jk}(x_j), \tag{36}$$

which, once evaluated in P_0, allow to determine 6ω components of y_α. It is worth observing that, since the initial conditions can be assigned arbitrarily, such 6ω quantities can assume arbitrary values. Moreover, there are further 3ω components of the vector y_α that remain completely arbitrary, since the system (34) and the initial relations (36) allow to determine only 7ω of the 10ω components of y_α. It is not guaranteed that the Inequality (35) is satisfied whatever is $y_\alpha \in H_0$. Thus, we define the space $W_0 \subseteq H_0$ constituted by the vectors of H_0, which satisfy both Equation (34) and the Inequality (35).

Remark 4. *It is worth observing that, although it is not guaranteed that the Inequality (35) is satisfied, whatever is $y_\alpha \in H_0$, at this stage, we do not have elements to exclude such a possibility. In other words, we do not have elements to decide if, actually, W_0 is a proper subspace of H_0 or if it coincides with H_0.*

In order to decide if $W_0 \subset H_0$, or $W_0 = H_0$, we follow the way paved by Muschik and Ehrentraut [13], who observed that such a decision cannot be ensued solely by the second law of thermodynamics, because such a law does not contain information regarding Equation (34) or the initial conditions (36). In order to fill this gap, Muschik and Ehrentraut completed the information contained in the Inequality (35) by an amendment that clarifies how the reversible transformations can be realized from the operative point of view. Here, we followed their strategy, but we propose a more general approach that includes the amendment into a new formulation of the second law. To achieve that task, we needed some preliminary definitions. To this end, we observed that, in the real world, reversible thermodynamic transformations do not exist, but they are approximated by very slow (quasi-static) transformations in which, in any point $P_0 \in B$, the system is close to the thermodynamic equilibrium. From an ideal point of view, a quasi-static transformation requires an infinite time to occur, and in any point of the system, the value of the state variable is constant in time.

Remark 5. *Alternative formulations of the thermodynamic laws that consider realistic transformations occurring in a finite time are proposed within the framework of finite time thermodynamics [16–18].*

As far as the thermodynamic framework is concerned, if B undergoes a quasi-static transformation, along with Muschik and Ehrentraut [13], we say that, in any point (P_0, t_0), the vectors of the higher derivatives are elements of E_0. Such an observation suggests the following definitions.

Definition 11. *A vector $y_\alpha \in H_0$ is said:*

- *Real, if it satisfies the relation $B_\alpha(\Sigma_0)y_\alpha > D(\Sigma_0)$;*
- *Ideal, if it satisfies the relation $B_\alpha(\Sigma_0)y_\alpha = D(\Sigma_0)$;*
- *Over-ideal, if it satisfies the relation $B_\alpha(\Sigma_0)y_\alpha < D(\Sigma_0)$.*

We can establish the following, owing to the definitions above:

Postulate 1 (A local formulation of the second law of thermodynamics). *Let B be a body, and let the couple (P_0, t_0) represent an arbitrary point of B at an arbitrary instant $t_0 \in [\tau_0, \tau_0 + \tau]$. Suppose B undergoes an arbitrary thermodynamic process of initial instant t_0 and duration $\tau_0 + \tau - t_0$. Then, the local space of the higher derivatives W_0 does not contain over-ideal vectors. Moreover, a vector $y_\alpha \in W_0$ is ideal if, and only if, (P_0, t_0) is in thermodynamic equilibrium.*

The postulate, the above traduces the experimental evidence, in that, in a thermo-dynamic process, the entropy production cannot be negative at any point P_0 of B at any instant t_0. Moreover, it also expresses the further experimental fact, which is often tacit in the formulations of the second law of thermodynamics—that the entropy production can be zero only in the points of B that are in equilibrium. In particular, if the point $P_0 \in B$ at the instant t_0 is in a thermodynamic equilibrium, then $W_0 = E_0$. Hence, $y_\alpha \in E_0$ is ideal if, and only if, (P_0, t_0) is in a thermodynamic equilibrium. Finally, a vector $y_\alpha \in W_0$ is either real or over-ideal if, and only if, (P_0, t_0) is not in a thermodynamic equilibrium.

Remark 6. *We should note that the local formulation of the second law of thermodynamics prohibits that over-ideal vectors be in W_0, but it does not prevent that they be in H_0. Whether H_0 contains over-ideal vectors (or not) is the focus of the present investigation.*

Definition 12. *Let B be a body undergoing an arbitrary thermodynamic process p of the initial instant t_0 and duration $\tau_0 + \tau - t_0$, and let γ be the curve representative of the process in C. The process p is said to be:*

- *Irreversible, if there exists at least a point $z_\alpha(P, t)$ of γ in which the vector of the higher derivatives $y_\alpha(P, t)$ is real;*

- *Reversible, if in any point $z_\alpha(P, t)$ of γ the vector of the higher derivatives $y_\alpha(P, t)$ is ideal;*

- *Over-reversible, if there exists at least a point $z_\alpha(P, t)$ of γ in which the vector of the higher derivatives $y_\alpha(P, t)$ is over-ideal.*

The definitions above allow us to enunciate the following:

Postulate 2 (The global formulation of the second law of thermodynamics). *Over-reversible processes do not occur in nature. Moreover, a thermodynamic process is reversible if, and only if, any point $P \in B$, at any instant t, is in a thermodynamic equilibrium.*

The previous formulations (local and global) of the second law of thermodynamics include information not present in the classical ones—that the reversible transformations are necessarily quasi-static and, hence, they need an infinite time to occur. Thus, they represent ideal processes, which, in nature, are approximated by very slow transformations. Here, we take into account such a situation by admitting that this is possible if, and only if, at any point of the body, at any instant t, there is a state of thermodynamic equilibrium. On the other hand, this implies that at any point of γ, the vector of the higher derivatives $y_\alpha(P, t)$ lie in the equilibrium bundle $\mathcal{E}$, and is ideal.

4. The Muschik and Ehrentraut Theorem Revisited

In this section, we present a novel formulation of the Muschik and Ehrentraut theorem proved in reference [13]. To this end, we use the thermodynamic framework and the generalized formulations of the second law established above.

Theorem 1. *Let B be a body, and let the couple (P_0, t_0) represent an arbitrary point of B at an arbitrary instant $t_0 \in [\tau_0, \tau_0 + \tau]$. Then, $H_0 = W_0$.*

Proof. To prove the theorem, it is enough to demonstrate that the vectors of H_0 are all (and only) the vectors of W_0. To this end, we observe that, in the generic point (P_0, t_0), fixed values of $A_{\beta\alpha}(\Sigma_0)$, $C_\beta(\Sigma_0)$, $B_\alpha(\Sigma_0)$, and $D(\Sigma_0)$ correspond to infinite vectors $y_\alpha(P_0, t_0)$, because only ω components of $y_\alpha(P_0, t_0)$ are determined by the balance equations while the remaining 9ω are completely arbitrary (see discussion in Section 3). Moreover, if all the y_α in H_0 are over-ideal, the vector space W_0 would be empty, because the second law of thermodynamics prohibits that it contains over-ideal vectors. As a consequence, in (P_0, t_0), no process would be possible. On the other hand, since (P_0, t_0) is arbitrary, no thermodynamic transformation could occur in B in the interval of time $[\tau_0, \tau_0 + \tau]$. So, in (P_0, t_0), the space H_0 contains, in principle, both real/ideal vectors and over-ideal ones.

Let us suppose that in (P_0, t_0), the space H_0 contains an ideal vector y_α^1 and an over-ideal vector y_α^2. Since the existence of y_α^1 is possible if, and only if, (P_0, t_0) is in the thermodynamic equilibrium, while y_α^2 exists if, and only if (P_0, t_0) is not in the thermodynamic equilibrium, such a situation is impossible to be realized.

Analogously, let us suppose that y_α^1 is ideal and y_α^2 is real. Again, such a situation is impossible, because it would require (P_0, t_0) to be in equilibrium and not in equilibrium.

Finally, let y_α^1 be a real vector, and y_α^2 be an over-ideal one. Such a situation is possible, in principle, provided (P_0, t_0) is not in equilibrium.

In such a case, due to the local formulation of the second law, neither y_α^1 nor y_α^2 are elements of E_0.

Let us consider the linear combination $y_\alpha^3 = \lambda y_\alpha^1 + (1 - \lambda)y_\alpha^2$, with $\lambda \in]0, 1[$. Since y_α^1 and y_α^2 are in H_0, they satisfy the following equations

$$A_{\beta\alpha}(\Sigma_0)y_\alpha^1 = C_\beta(\Sigma_0),\tag{37}$$

$$A_{\beta\alpha}(\Sigma_0)y_\alpha^2 = C_\beta(\Sigma_0).\tag{38}$$

The combination of Equation (37) multiplied by λ and Equation (38) multiplied by $(1 - \lambda)$ leads to

$$A_{\beta\alpha}(\Sigma_0)y_\alpha^3 = C_\beta(\Sigma_0),\tag{39}$$

namely, y_α^3 is also a solution of Equation (34), i.e., it is in H_0. On the other hand, the local entropy production corresponding to y_α^3 can be written as

$$\sigma^3 = \lambda\Big[B_\alpha(\Sigma_0)y_\alpha^1 - D(\Sigma_0)\Big] + (1 - \lambda)\Big[B_\alpha(\Sigma_0)y_\alpha^2 - D(\Sigma_0)\Big] =\tag{40}$$

$$= B_\alpha(\Sigma_0)\Big[\lambda y_\alpha^1 + (1 - \lambda)y_\alpha^2\Big] - D(\Sigma_0).$$

Since λ is arbitrary in $]0, 1[$, nothing prevents choosing it as

$$\lambda = \frac{D(\Sigma_0) - B_\alpha(\Sigma_0)y_\alpha^2}{B_\alpha(\Sigma_0)[y_\alpha^1 - y_\alpha^2]},\tag{41}$$

because, as it is easily seen, the right-hand side of Equation (41) is in the interval $]0, 1[$. In fact, being y_α^2 over-ideal we have $D(\Sigma_0) - B_\alpha(\Sigma_0)y_\alpha^2 > 0$. Moreover, being y_α^1 real, we have $B_\alpha(\Sigma_0)[y_\alpha^1 - y_\alpha^2] > D(\Sigma_0) - D(\Sigma_0)$, namely, $B_\alpha(\Sigma_0)[y_\alpha^1 - y_\alpha^2] > 0$. Hence, $\lambda > 0$. Moreover, being y_α^1 real, we also have $B_\alpha(\Sigma_0)[y_\alpha^1 - y_\alpha^2] > D(\Sigma_0) - B_\alpha(\Sigma_0)y_\alpha^2$ and, hence, $\lambda < 1$.

Consequently, the right-hand side of Equation (40) vanishes, so that y_α^3 is in E_0. However, this is impossible, otherwise (P_0, t_0) would be in a thermodynamic equilibrium. Thus, it is forbidden that in (P_0, t_0) there are both real and over-ideal vectors that are solutions of the local balance laws (34).

Furthermore, suppose that both y_α^1 and y_α^2 are real. Then, it is easy to verify by direct calculations that λ can be taken, such that $\sigma^3 > 0$.

Finally, if (P_0, t_0) is a point of equilibrium, then the entropy production related to y_α^1 and y_α^2 vanishes, so that by Equation (40), it follows that σ^3 is zero.

The considerations above show the impossibility that, in a point P_0 of B, at a given instant t_0, the solutions of Equation (34) can be of different types. Moreover, they cannot be over-ideal only, because this contradicts the local form of the second law of thermodynamics.

Thus, H_0 may contain either only real vectors, and in such a case, (P_0, t_0) is a point of non-equilibrium, or only ideal vectors, and in such a case, (P_0, t_0) is a point of equilibrium. This conclusion proves the theorem. $\square$

Corollary 2. $\mathcal{H} = \mathcal{W}$.

Proof. This corollary is an immediate consequence of the arbitrariness of the initial instant t_0, and of the point P_0. In particular, whatever t_0 is, we can ever consider it as the initial instant of the restricted process of duration $\tau_0 + \tau - t_0$, so that $H_0 \equiv H_t(P_0, t_0)$ has dimension 10ω. Moreover, only 7ω of components of the vectors of H_0 can be determined by the algebraic relations (34) and (36) while the further 3ω components are completely arbitrary. Thus, to H_0 can be applied to the conclusions established in Theorem 1. This is enough to prove that, for any $t \in [\tau_0, \tau_0 + \tau]$, the space of the higher derivatives H_t contains only real or ideal vectors. $\square$

Remark 7. *The Corollary* 2 *also implies* $\hat{\mathcal{E}} = \mathcal{E}$.

Corollary 3. *The unilateral differential constraint* (32) *is a restriction on the constitutive quantities* U_β, r_β, s, *and* J_k, *and not on the thermodynamic processes* p.

Proof. In fact, any process $p : t \in [\tau_0, \tau_0 + \tau] \subseteq \mathbb{R} \to z_\alpha(x_j, t) \in \mathcal{C}$, where $z_\alpha(x_j, t)$ is a solution of the balance laws (29), can only be either irreversible or reversible but not over-reversible, because otherwise its representative curve γ would contain at least an over-ideal point against Corollary 2. On the other hand, such a property of the solutions of the system of balance laws is not guaranteed, whatever $A_{\beta\alpha}$ and C_β are, and for arbitrary s and J_k, because, given the state space, only particular forms of those functions defined on it lead to a nonnegative entropy production. Then, the role of the unilateral differential constraint in Equation (32) is just to select such forms. $\square$

5. Discussion

Exploitation of the second law of thermodynamics is based on the assumption that the constitutive equations modeling material properties must be assigned in such a way that all solutions of the field equations satisfy the entropy inequality. An alternative interpretation of the second law consequences is that we must exclude from the set of solutions of the balance equations ones that do not guarantee a nonnegative entropy production. The problem of the choice between the two interpretations above was solved in 1996 by Muschik and Ehrentraut [13], by postulating an amendment to the second law, which states that, at a fixed instant of time and in any point of the body, the entropy production is zero if, and only if, this point is in a thermodynamic equilibrium. Muschik and Ehrentraut proved that, under the validity of the amendment, the constitutive equations and not the thermodynamic processes are restricted by the second law of thermodynamics. Such a result provides the theoretical basis to the rigorous mathematical procedures for the exploitation of the second law proposed by Coleman and Noll in 1963 [1], and by Liu in 1972 [12].

In the present paper, we revisited the Muschik and Ehrentraut theorem, highlighting some geometric aspects that were hidden in reference [13]. Moreover, we proposed a generalized formulation of the second law of thermodynamics that incorporates the amendment. Progresses in the analysis of the entropy principle achieved in this way are the following:

1. In the analysis of the efficiency of the thermodynamic systems, the concept of the Carnot cycle plays a fundamental role. Carnot was the first to prove that the efficiency of a quasi-static transformation is maximum for a Carnot cycle. However, since quasi-static transformations require an infinite time, the Carnot efficiency is not suitable to describe the performance of real heat engines, which produce irreversible transformations taking over in a finite time. Thus, for real processes, it is important to investigate how much the efficiency deteriorates when the cycle is operated in a

finite time. This is the task of finite time thermodynamics, a modern non-equilibrium theory, which has been developed in the last four decades [16–18]. Since the global formulation of the second law of thermodynamics proposed here explicitly mentions the fact that a reversible transformation lies in the equilibrium bundle $\mathcal{E}$, it seems to be more appropriate for finite time thermodynamics with respect to the classical formulation, which does not contain such information.

2. Owing to the Definitions 11 and 12, we have given a general but intuitive definition of reversible and irreversible processes, by analyzing the properties of its representative curve in the fiber bundle of the configuration spaces. To the best of our knowledge, such a definition has never been presented in literature. Furthermore, such properties provide a generalized formulation, both local and global, of the second law of thermodynamics, leading to the main results of the present paper, namely, Theorem 1.

3. As a final consideration, we observe that, in non-equilibrium thermodynamics, the concepts of entropy and temperature are still open problems, since their definitions, far from equilibrium, are questionable [6]. Usually, the problem is solved by postulating the principle of the local equilibrium, which assumes that, at least locally, the thermodynamic functions defined at the equilibrium can describe the thermodynamic properties of the materials in non-equilibrium situations [19]. Thus, our geometric framework, providing a meaningful representation of the state spaces involved in equilibrium and non-equilibrium processes, allows to interpret the validity of the principle of the local equilibrium as the extension of some properties that are valid in the subspace $\mathcal{E}$ of $\mathcal{H}$, to the whole $\mathcal{H}$. In Figure 1, we provide a sketch of the analysis of the entropy principle developed in the present paper and its relation with the Muschik and Ehrentraut theory.

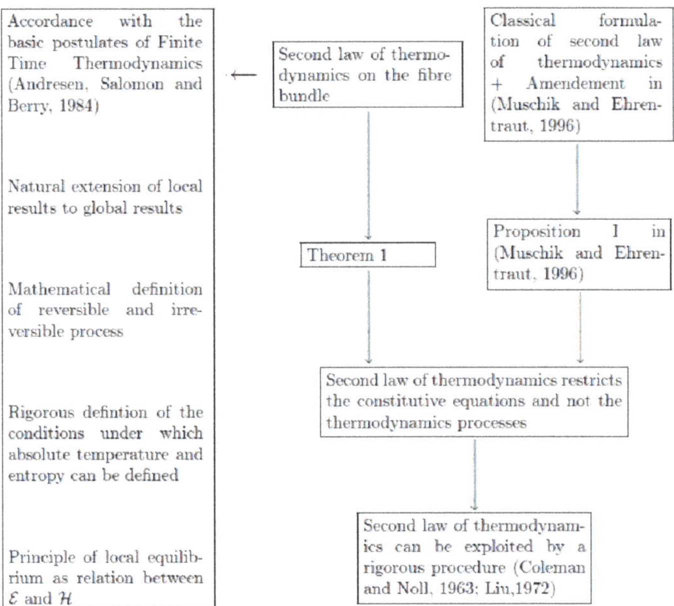

Figure 1. A schematized explanation of the analysis of the entropy principle developed in the present paper, and of its relation to the Muschik and Ehrentraut theory. An Amendment to second law of thermodynamics has been proposed by Muschik and Ehrentraut in 1996 [13]; rigorous procedures for the exploitation of the entropy principle have been formulated by Coleman and Noll in 1963 [1], and by Liu in 1972 [12]; the general tenets of Finite Time Thermodynamics have been summarized by Andresen, Salomon and Berry in 1984 [17].

In reference [21], the results in [13] were extended in order to encompass non-regular processes. Such an investigation deserves consideration, since thermodynamic processes involving discontinuous solutions are frequent in physics. In future research studies, we will aim to reanalyze the results in [21], within the mathematical framework presented here.

Author Contributions: Conceptualization, V.A.C. and P.R.; formal analysis, V.A.C. and P.R.; investigation, V.A.C. and P.R.; writing—original draft, V.A.C. and P.R.; writing—review and editing, V.A.C. and P.R. All authors have read and agreed to the published version of the manuscript.

Funding: This research was funded by the University of Basilicata (RIL 2020), the University of Messina (FFABR Unime 2020), and the Italian National Group of Mathematical Physics (GNFM-INdAM).

Institutional Review Board Statement: Not applicable.

Informed Consent Statement: Not applicable.

Data Availability Statement: Not applicable.

Acknowledgments: Patrizia Rogolino thanks the University of Messina and the Italian National Group of Mathematical Physics (GNFM-INdAM) for financial support. Vito Antonio Cimmelli thanks the University of Basilicata and the Italian National Group of Mathematical Physics (GNFM-INdAM) for financial support.

Conflicts of Interest: The authors declare no conflict of interest.

References

1. Coleman, B.D.; Noll, W. The thermodynamics of elastic materials with heat conduction and viscosity. *Arch. Ration. Mech. Anal.* **1963**, *13*, 167–178. [CrossRef]
2. Truesdell, C. *Rational Thermodynamics*, 2nd ed.; Springer: New York, NY, USA, 1984.
3. Grad, H. On the kinetic theory of rarefied gases. *Commun. Pure Appl. Math.* **1949**, *2*, 331–407 [CrossRef]
4. Jou, D.; Casas-Vázquez, J.; Lebon, G. *Extended Irreversible Thermodynamics*, 4th ed.; Springer: Berlin, Germany, 2010.
5. Müller, I.; Ruggeri, T. *Rational Extended Thermodynamics*, 2nd ed.; Springer: New York, NY, USA, 1998.
6. Sellitto, A.; Cimmelli, V.A.; Jou, D. *Mesoscopic Theories of Heat Transport in Nanosystems*; Springer: Berlin, Germany, 2016.
7. Szucs, M.; Kovacs, R.; Simic, S. Open Mathematical Aspects of Continuum Thermodynamics: Hyperbolicity, Boundaries and Nonlinearities. *Symmetry* **2020**, *12*, 1469. [CrossRef]
8. Smith, G.F. On isotropic functions of symmetric tensors, skew-symmetric tensors and vectors. *Int. J. Eng. Sci.* **1971**, *9*, 899–916. [CrossRef]
9. Oliveri, F. Lie Symmetries of Differential Equations: Classical Results and Recent Contributions. *Symmetry* **2010**, *2*, 658–706. [CrossRef]
10. Cimmelli, V.A.; Jou, D.; Ruggeri, T.; Ván, P. Entropy Principle and Recent Results in Non-Equilibrium Theories. *Entropy* **2014**, *16*, 1756–1807. [CrossRef]
11. Triani, V.; Papenfuss, C.; Cimmelli, V.A.; Muschik, W. Exploitation of the Second Law: Coleman-Noll and Liu Procedure in Comparison. *J. Non-Equilib. Thermodyn.* **2008**, *33*, 47–60. [CrossRef]
12. Liu, I.-S. Method of Lagrange multipliers for exploitation of the entropy principle. *Arch. Ration. Mech. Anal.* **1972**, *46*, 131–148. [CrossRef]
13. Muschik, W.; Ehrentraut, H. An Amendment to the Second Law. *J. Non-Equilib. Thermodyn.* **1996**, *21*, 175–192. [CrossRef]
14. Dolfin, M.; Francaviglia, M.; Rogolino, P. A Geometric Perspective on Irreversible Thermodynamics with Internal Variables. *J. Non-Equilib. Thermodyn.* **1998**, *23*, 250–263. [CrossRef]
15. Dolfin, M.; Francaviglia, M.; Rogolino, P. A geometric model for the thermodynamics of simple materials. *Period. Polytech. Mech. Eng.* **1999**, *43*, 29–37.
16. Andresen, B.; Berry, R.S.; Nitzan, A.; Salomon, P. Thermodynamics in finite time. 1. The step-Carnot cicle. *Phys. Rev. A* **1977**, *15*, 2086–2093. [CrossRef]
17. Andresen, B.; Salomon, P.; Berry, R.S. Thermodynamics in finite time. *Phys. Today* **1984**, *37*, 62–70. [CrossRef]
18. Hoffmann, K.H. Recent Developments in Finite Time Thermodynamics. *Tech. Mech.* **2002**, *22*, 14–25.
19. De Groot, S.R.; Mazur, P. *Nonequilibrium Thermodynamics*; North Holland Publishing Company: Amsterdam, The Netherlands, 1962.
20. Courant, R.; Hilbert, D. *Methods of Mathematical Physics: Partial Differential Equations*; John Wiley and Sons: New York, NY, USA, 1989; Volume II.
21. Triani, V.; Cimmelli, V.A. Interpretation of Second Law of Thermodynamics in the presence of interfaces. *Contin. Mech. Thermodyn.* **2012**, *24*, 165–174. [CrossRef]

Article

Integrability of the Multi-Species TASEP with Species-Dependent Rates

Eunghyun Lee

Department of Mathematics, Nazarbayev University, Nur-Sultan 010000, Kazakhstan; eunghyun.lee@nu.edu.kz

Abstract: Assume that each species l has its own jump rate b_l in the multi-species totally asymmetric simple exclusion process. We show that this model is *integrable* in the sense that the Bethe ansatz method is applicable to obtain the transition probabilities for all possible N-particle systems with up to N different species.

Keywords: TASEP; ASEP; multi-species; Bethe ansatz

Citation: Lee, E. Integrability of the Multi-Species TASEP with Species-Dependent Rates. *Symmetry* **2021**, *13*, 1578. https://doi.org/10.3390/sym13091578

Academic Editors: Róbert Kovács, Patrizia Rogolino, Francesco Oliveri and Jaume Giné

Received: 28 July 2021
Accepted: 21 August 2021
Published: 27 August 2021

Publisher's Note: MDPI stays neutral with regard to jurisdictional claims in published maps and institutional affiliations.

1. Introduction

The multi-species asymmetric simple exclusion process on $\mathbb{Z}$ is a generalization of the asymmetric simple exclusion process (ASEP) on $\mathbb{Z}$ in the sense that each particle may belong to a different species labelled by an integer $l \in \{1, 2, \cdots\}$. Each particle jumps to the right by one step with the probability p or to the left by one step with the probability $q = 1 - p$ after a waiting time that is exponentially distributed with rate 1. If a particle belonging to l attempts to jump to the site occupied by a particle belonging to $l' \geq l$, the jump is prohibited; however, if a particle belonging to l' attempts to jump to the site occupied by a particle belonging to $l < l'$, then the jump occurs by interchanging positions.

The transition probabilities and some determinantal formulas for the multi-species ASEP and its special cases were found in [1–5]. For certain special initial conditions with a single second class particle, some distributions and their asymptotics were studied in [6,7]. More recently, asymptotic behaviors of the second class particles were studied using the color-position symmetry, see [8]. In fact, the multi-species asymmetric simple exclusion process can be considered in a more general context—that is, the coloured six vertex model [9].

Another direction of generalizing the ASEP and other models studied in the integrable probability is to make the jump rates inhomogeneous. It is known that the Bethe ansatz method is still applicable to some single-species model with inhomogeneous jump rates. The basic idea of using the Bethe ansatz in the ASEP is that the generator of the ASEP is a similarity transformation of the XXZ quantum spin system. Considering that the Bethe ansatz is a method to find eigenvalues and eigenvectors of a certain class of quantum spin systems, we use the Bethe ansatz to find the solution of the forward equation of a certain class of Markov processes, that is, the transition probabilities of the processes.

Of course, for some particle models, the Bethe ansatz method cannot be used. For the background of Bethe ansatz, see [10,11]. It is known that the Bethe ansatz is applicable to some generalization of the ASEP. For example, the transition probability and the current distribution of the totally asymmetric simple exclusion process (TASEP) with particle-dependent rates were studied in [12], and the transition probabilities and some asymptotic results for the q deformed totally asymmetric zero range process with site-dependent rates were studied in [13,14].

In this paper, we consider the multi-species totally asymmetric simple exclusion processes with N particles in which particles move to the right and each species l is allowed to have its own rate b_l. Following the notations used in [4], let $X = (x_1, \cdots, x_N) \in \mathbb{Z}^N$ with $x_1 < \cdots < x_N$ represent the positions of particles, and let $\pi = \pi(1)\pi(2)\cdots\pi(N)$

be a permutation of a multi-set $\mathcal{M} = [i_1, \cdots, i_N]$ with elements taken from $\{1, \cdots, N\}$ and $\pi(i)$ representing the species of the ith leftmost particle. Then, the state of an N-particle system is denoted by

$$(X, \pi) = (x_1, \cdots, x_N, \pi(1), \cdots, \pi(N)).$$

Let us write $P_{(Y,v)}(X, \pi; t)$ for the transition probability from (Y, v) at $t = 0$ to (X, π) at a later time t. For fixed X and Y, $P_{(Y,v)}(X, \pi; t)$ is regarded as a matrix element of an $N^N \times N^N$ matrix denoted by $\mathbf{P}_Y(X; t)$ whose columns and rows are labelled by $v, \pi = 1 \cdots 11, 1 \cdots 12, \cdots, N \cdots N$, respectively. Throughout this paper, given an $N^n \times N^n$ matrix, we assume that its rows $(i_1 \cdots i_n)$ and columns $(j_1 \cdots j_n)$ are labelled by $1 \cdots 1, \cdots, n \cdots n$ and that these labels are listed lexicographically, unless stated otherwise. The main result of this paper is that the multi-species TASEP with species-dependent rates is an integrable model, and we provide a formula analogous to (2.12) in [4] using the Bethe ansatz method.

Statement of the Results

We first introduce a few objects to state the main theorem. Define an $N^2 \times N^2$ matrix $\mathbf{R}_{\beta\alpha} = \left[R_{ij,kl} \right]$ with

$$R_{ij,kl} = \begin{cases} S_{\beta\alpha}(i) & \text{if } ij = kl \text{ with } i \leq j; \\ -1 & \text{if } ij = kl \text{ with } i > j; \\ T_{\beta\alpha}(i) & \text{if } ij = lk \text{ with } i < j; \\ 0 & \text{for all other cases,} \end{cases} \tag{1}$$

where

$$S_{\beta\alpha}(i) = -\frac{1 - b_i \xi_\beta}{1 - b_i \xi_\alpha}, \quad T_{\beta\alpha}(i) = \frac{b_i(\xi_\beta - \xi_\alpha)}{1 - b_i \xi_\alpha}, \quad \xi_\alpha, \xi_\beta \in \mathbb{C}. \tag{2}$$

Remark 1. *The form of the matrix* (1) *was obtained by induction via similar arguments to Sections 2.1 and 2.2 in* [3]*, which treats a special case, and the motivation of* (2) *is given in Section 2.1. Finding the form of* (1) *with* (2) *is the key idea of this paper.*

Let $\mathcal{T}_l$ be the simple transposition that interchanges the number at the lth slot and the number at the $(l+1)$st slot. If $\mathcal{T}_l$ maps a permutation $(\ \cdots \alpha\beta \cdots\ )$ to $(\ \cdots \beta\alpha \cdots\ )$, we write $\mathcal{T}_l = \mathcal{T}_l(\beta, \alpha)$ when necessary. Corresponding to a simple transposition $\mathcal{T}_l(\beta, \alpha)$, we define $N^N \times N^N$ matrix $\mathbf{T}_l(\beta, \alpha)$ by the tensor product of matrices,

$$\mathbf{T}_l(\beta, \alpha) = \underbrace{\mathbf{I}_N \otimes \cdots \otimes \mathbf{I}_N}_{(l-1) \text{ times}} \otimes \mathbf{R}_{\beta\alpha} \otimes \underbrace{\mathbf{I}_N \otimes \cdots \otimes \mathbf{I}_N}_{(N-l-1) \text{ times}} \tag{3}$$

where $\mathbf{I}_N$ is the $N \times N$ identity matrix. For a permutation σ in the symmetric group $\mathcal{S}_N$ written

$$\sigma = \mathcal{T}_{i_j} \cdots \mathcal{T}_{i_1} = \mathcal{T}_{i_j}(\beta_j, \alpha_j) \cdots \mathcal{T}_{i_1}(\beta_1, \alpha_1) \tag{4}$$

for some $i_1, \cdots, i_j \in \{1, \cdots, N-1\}$, we define

$$\mathbf{A}_\sigma = \mathbf{T}_{i_j}(\beta_j, \alpha_j) \cdots \mathbf{T}_{i_1}(\beta_1, \alpha_1). \tag{5}$$

Here, $\mathbf{A}_\sigma$ is well defined, that is, $\mathbf{A}_\sigma$ is unique regardless of the representation of σ by simple transpositions. This *well-definedness* is due to the following lemma.

Lemma 1. *The following consistency relations are satisfied.*
(a) $\ \mathbf{T}_i(\beta, \alpha)\mathbf{T}_j(\delta, \gamma) = \mathbf{T}_j(\delta, \gamma)\mathbf{T}_i(\beta, \alpha) \quad \text{if } |i - j| \geq 2.$
(b) $\ \mathbf{T}_i(\gamma, \beta)\mathbf{T}_j(\gamma, \alpha)\mathbf{T}_i(\beta, \alpha) = \mathbf{T}_j(\beta, \alpha)\mathbf{T}_i(\gamma, \alpha)\mathbf{T}_j(\gamma, \beta) \quad \text{if } |i - j| = 1.$
(c) $\ \mathbf{T}_i(\beta, \alpha)\mathbf{T}_i(\alpha, \beta) = \mathbf{I}_{NN}.$

The relations in Lemma 1 with $b_l = 1$ for all l are already known for the multi-species ASEP.

Remark 2. *The definitions of $\mathcal{T}_l$ and $\mathbf{A}_\sigma$ are motivated by the arguments for $N = 2, 3$ in Sections 2.1 and 2.2 in [3], which treats a special case.*

Let $\mathbf{J}(t)$ be the $N^N \times N^N$ diagonal matrix whose (π, π)-element is given by $e^{\varepsilon_{\pi\pi} t}$ where

$$\varepsilon_{\pi\pi} = \varepsilon_{\pi\pi}(\xi_1, \cdots, \xi_N) = \frac{1}{\xi_1} + \cdots + \frac{1}{\xi_N} - \left(b_{\pi(1)} + \cdots + b_{\pi(N)} \right), \quad \xi_1, \cdots, \xi_N \in \mathbb{C},$$

let $\mathbf{D}(x_1, \cdots, x_N)$ be the $N^N \times N^N$ diagonal matrix whose (π, π)-element is given by $b_{\pi(1)}^{x_1} \cdots b_{\pi(N)}^{x_N}$, and let $\mathbf{D}'(y_1, \cdots, y_N)$ be the $N^N \times N^N$ diagonal matrix whose (ν, ν)-element is given by $b_{\nu(1)}^{-y_1} \cdots b_{\nu(N)}^{-y_N}$ where y_is are the initial positions. In the next theorem, the integral of a matrix implies that the integral is taken element-wise, and $\oint$ implies $\frac{1}{2\pi i} \int$.

Theorem 1. *Let $\mathbf{A}_\sigma$ be given as in (5) and c be a positively oriented circle centred at the origin with a radius less than b_l for all l in the complex plane $\mathbb{C}$. Then, the matrix of the transition probabilities of the multi-species TASEP with species-dependent rates is*

$$\mathbf{P}_Y(X; t) = \sum_{\sigma \in \mathcal{S}_N} \oint_c \cdots \oint_c \mathbf{J}(t) \mathbf{D}(x_1, \cdots, x_N) \mathbf{A}_\sigma \mathbf{D}'(y_1, \cdots, y_N) \prod_{i=1}^{N} \left(\xi_{\sigma(i)}^{x_i - y_{\sigma(i)} - 1} \right) d\xi_1 \cdots d\xi_N. \quad (6)$$

Remark 3. *If $\nu = 12 \cdots N$, in other words, all N particles belong to different species, and the species are initially arranged in ascending order, then the system is the same as the TASEP with particle-dependent rates studied in [12]. Hence, the transition probability $P_{(Y, 12 \cdots N)}(X, 12 \cdots N; t)$ can be expressed as a determinant (see Theorem 1 in [12]).*

Remark 4. *Theorem 1 partially extends (2.12) in [4]. In other words, (6) with $b_l = 1$ is equal to (2.12) in [4] with $p = 1$.*

The proofs of Lemma 1 and Theorem 1 are given in the next section.

2. Proof of Theorem 1

In order to prove that the (π, ν)-element of the right-hand side of (6) is $P_{(Y,\nu)}(X, \pi; t)$, we should show that the (π, ν)-element satisfies its forward equation and the initial condition $P_{(Y,\nu)}(X, \pi; 0) = \delta_{X,Y}$.

2.1. Forward Equations

We first study the two-particle systems, which will be building-blocks for the formulas for N-particle systems. When $x_1 < x_2 - 1$, the forward equations of $P_{(Y,\nu)}(x_1, x_2, \pi; t)$ are straightforward because two particles act as *free* particles. Hence, the forward equations of $P_{(Y,\nu)}(x_1, x_2, \pi; t)$ are expressed as

$$\frac{d}{dt} \mathbf{P}_Y(x_1, x_2; t) = \underbrace{\begin{bmatrix} b_1 & 0 & 0 & 0 \\ 0 & b_1 & 0 & 0 \\ 0 & 0 & b_2 & 0 \\ 0 & 0 & 0 & b_2 \end{bmatrix}}_{(r_1)} \mathbf{P}_Y(x_1 - 1, x_2; t) + \underbrace{\begin{bmatrix} b_1 & 0 & 0 & 0 \\ 0 & b_2 & 0 & 0 \\ 0 & 0 & b_1 & 0 \\ 0 & 0 & 0 & b_2 \end{bmatrix}}_{(r_2)} \mathbf{P}_Y(x_1, x_2 - 1; t)$$

$$- \begin{bmatrix} b_1 + b_1 & 0 & 0 & 0 \\ 0 & b_1 + b_2 & 0 & 0 \\ 0 & 0 & b_2 + b_1 & 0 \\ 0 & 0 & 0 & b_2 + b_2 \end{bmatrix} \mathbf{P}_Y(x_1, x_2; t) \quad (7)$$

where the derivative of the matrix $\mathbf{P}_Y(x_1, x_2; t)$ on the left-hand side implies the matrix of the derivatives of elements of $\mathbf{P}_Y(x_1, x_2; t)$. The matrices (r_1) and (r_2) account for the *probability current* in the states (x_1, x_2, π) by a particle's jump to the right next site, which is empty. On the other hand, when $x_1 = x_2 - 1$, if two particles belong to different species, two particles may swap their positions. For example, if the initial state is $(Y, 12)$, the system cannot be at $(X, 21)$ at any later time t. Hence, the forward equation of $P_{(Y,12)}(x_1, x_1 + 1, 12; t)$ is

$$\frac{d}{dt} P_{(Y,12)}(x_1, x_1 + 1, 12; t) = b_1 P_{(Y,12)}(x_1 - 1, x_1 + 1, 12; t) - b_2 P_{(Y,12)}(x_1, x_1 + 1, 12; t).$$

On the other hand, $P_{(Y,12)}(x_1, x_1 + 1, 21; t) = 0$ for all t, because the model is totally asymmetric. If the initial state is $(Y, 21)$, the forward equation of $P_{(Y,21)}(x_1, x_1 + 1, 21; t)$ is

$$\frac{d}{dt} P_{(Y,21)}(x_1, x_1 + 1, 21; t) = b_2 P_{(Y,21)}(x_1 - 1, x_1 + 1, 21; t) - (b_2 + b_1) P_{(Y,21)}(x_1, x_1 + 1, 21; t)$$

and the forward equation of $P_{(Y,21)}(x_1, x_1 + 1, 12; t)$ is

$$\frac{d}{dt} P_{(Y,21)}(x_1, x_1 + 1, 12; t) = b_1 P_{(Y,21)}(x_1 - 1, x_1 + 1, 12; t) + b_2 P_{(Y,21)}(x_1, x_1 + 1, 21; t)$$
$$- b_2 P_{(Y,21)}(x_1, x_1 + 1, 12; t).$$

Hence, the forward equations of $P_{(Y,\nu)}(x_1, x_1 + 1, \pi; t)$ are expressed as

$$\frac{d}{dt} \mathbf{P}_Y(x_1, x_1 + 1; t) =$$

$$\begin{bmatrix} b_1 & 0 & 0 & 0 \\ 0 & b_1 & 0 & 0 \\ 0 & 0 & b_2 & 0 \\ 0 & 0 & 0 & b_2 \end{bmatrix} \mathbf{P}_Y(x_1 - 1, x_1 + 1; t) + \underbrace{\begin{bmatrix} 0 & 0 & 0 & 0 \\ 0 & 0 & b_2 & 0 \\ 0 & 0 & 0 & 0 \\ 0 & 0 & 0 & 0 \end{bmatrix}}_{(B)} \mathbf{P}_Y(x_1, x_1 + 1; t)$$

$$- \left(\underbrace{\begin{bmatrix} 0 & 0 & 0 & 0 \\ 0 & 0 & 0 & 0 \\ 0 & 0 & b_2 & 0 \\ 0 & 0 & 0 & 0 \end{bmatrix}}_{(C)} + \begin{bmatrix} b_1 & 0 & 0 & 0 \\ 0 & b_2 & 0 & 0 \\ 0 & 0 & b_1 & 0 \\ 0 & 0 & 0 & b_2 \end{bmatrix} \right) \mathbf{P}_Y(x_1, x_1 + 1; t). \tag{8}$$

Here, the matrix (B) accounts for *probability current* going in the states $(x_1, x_1 + 1, 12)$ by the species-2 particle's jump from the state $(x_1, x_1 + 1, 21)$. Similarly, the matrix (C) accounts for *probability current* going out of the states $(x_1, x_1 + 1, 21)$ by species-2 particle's jump to the state $(x_1, x_1 + 1, 12)$. Equations (7) and (8) imply that, if $\mathbf{U}(x_1, x_2; t)$ is a 4×4 matrix whose elements are functions on $\mathbb{Z}^2 \times [0, \infty)$, then the forward equation of $P_{(Y,\nu)}(x_1, x_2, \pi; t)$ for any $x_1 < x_2$ is in the form of the (π, ν)-element of

$$\frac{d}{dt} \mathbf{U}(x_1, x_2; t) = \begin{bmatrix} b_1 & 0 & 0 & 0 \\ 0 & b_1 & 0 & 0 \\ 0 & 0 & b_2 & 0 \\ 0 & 0 & 0 & b_2 \end{bmatrix} \mathbf{U}(x_1 - 1, x_2; t) + \begin{bmatrix} b_1 & 0 & 0 & 0 \\ 0 & b_2 & 0 & 0 \\ 0 & 0 & b_1 & 0 \\ 0 & 0 & 0 & b_2 \end{bmatrix} \mathbf{U}(x_1, x_2 - 1; t)$$

$$- \begin{bmatrix} b_1 + b_1 & 0 & 0 & 0 \\ 0 & b_1 + b_2 & 0 & 0 \\ 0 & 0 & b_2 + b_1 & 0 \\ 0 & 0 & 0 & b_2 + b_2 \end{bmatrix} \mathbf{U}(x_1, x_2; t)$$

subject to the (π, ν)-element of

$$\begin{bmatrix} b_1 & 0 & 0 & 0 \\ 0 & b_2 & 0 & 0 \\ 0 & 0 & b_1 & 0 \\ 0 & 0 & 0 & b_2 \end{bmatrix} \mathbf{U}(x_1, x_1; t) = \begin{bmatrix} b_1 & 0 & 0 & 0 \\ 0 & b_1 & b_2 & 0 \\ 0 & 0 & 0 & 0 \\ 0 & 0 & 0 & b_2 \end{bmatrix} \mathbf{U}(x_1, x_1 + 1; t).$$

Now, we extend the argument for two-particle systems to N-particle systems. The matrices (r_1) and (r_2) in (7) for two-particle systems are generalized to

$$\mathbf{r}_i = \underbrace{\mathbf{I}_N \otimes \cdots \otimes \mathbf{I}_N}_{i-1} \otimes \mathbf{r} \otimes \underbrace{\mathbf{I}_N \otimes \cdots \otimes \mathbf{I}_N}_{N-i}, \quad i = 1, \cdots, N$$

where $\mathbf{r}$ is the diagonal matrix,

$$\mathbf{r} = \begin{bmatrix} b_1 & & \\ & \ddots & \\ & & b_N \end{bmatrix}.$$

The matrix (B) in (8) is generalized to an $N^2 \times N^2$ matrix $\mathbf{B} = \left[B_{ij,kl} \right]$ with

$$B_{ij,kl} = \begin{cases} b_j & \text{if } ij = lk \text{ with } i < j; \\ 0 & \text{for all other cases,} \end{cases}$$

and let

$$\mathbf{B}_i = \underbrace{\mathbf{I}_N \otimes \cdots \otimes \mathbf{I}_N}_{i-1} \otimes \mathbf{B} \otimes \underbrace{\mathbf{I}_N \otimes \cdots \otimes \mathbf{I}_N}_{N-i-1}.$$

The matrix (C) in (8) is generalized to $N^2 \times N^2$ matrix $\mathbf{C} = \left[C_{ij,kl} \right]$ with

$$C_{ij,kl} = \begin{cases} b_i & \text{if } ij = kl \text{ with } i < j; \\ 0 & \text{for all other cases,} \end{cases}$$

and let

$$\mathbf{C}_i = \underbrace{\mathbf{I}_N \otimes \cdots \otimes \mathbf{I}_N}_{i-1} \otimes \mathbf{C} \otimes \underbrace{\mathbf{I}_N \otimes \cdots \otimes \mathbf{I}_N}_{N-i-1}.$$

All forward equations of $P_{(Y,\nu)}(x_1, \cdots, x_N, \pi; t)$ may be expressed as a matrix equation. For example, if $x_i < x_{i+1} - 1$ for all i, then the forward equation of $P_{(Y,\nu)}(x_1, \cdots, x_N, \pi; t)$ is the (π, ν)-element of

$$\frac{d}{dt} \mathbf{P}_Y(x_1, \cdots, x_N; t) = \mathbf{r}_1 \mathbf{P}_Y(x_1 - 1, x_2, \cdots, x_N; t) + \cdots \tag{9}$$

$$+ \mathbf{r}_N \mathbf{P}_Y(x_1, \cdots, x_{N-1}, x_N - 1; t) - (\mathbf{r}_1 + \cdots + \mathbf{r}_N) \mathbf{P}_Y(x_1, \cdots, x_N; t)$$

and if $x = x_i = x_{i+1} - 1$ and $x_j < x_{j+1} - 1$ for all $j \neq i$,

$$\frac{d}{dt} \mathbf{P}_Y(x_1, \cdots, x_{i-1}, x, x+1, x_{i+2}, \cdots, x_N; t) =$$

$$+ \sum_{j \neq i+1} \mathbf{r}_j \mathbf{P}_Y(x_1, \cdots, x_{j-1}, x_j - 1, x_{j+1}, \cdots, x_N; t) + \mathbf{B}_i \mathbf{P}_Y(x_1, \cdots, x_{i-1}, x, x+1, x_{i+2}, \cdots, x_N; t) \tag{10}$$

$$- \sum_{j \neq i} \mathbf{r}_j \mathbf{P}_Y(x_1, \cdots, x_{j-1}, x, x+1, x_{j+2}, \cdots, x_N; t) - \mathbf{C}_i \mathbf{P}_Y(x_1, \cdots, x_{i-1}, x, x+1, x_{i+2}, \cdots, x_N; t).$$

For other configurations of $(x_1, \cdots, x_N)$, the form of the matrix of the forward equations may be different from (9) and (10). However, as in other Bethe ansatz applicable models, if $\mathbf{U}(x_1, \cdots, x_N; t) = \left[U_{\pi \nu}(x_1, \cdots, x_N; t) \right]$ is an $N^N \times N^N$ matrix whose elements $U_{\pi v}(x_1, \cdots, x_N; t)$ are functions on $\mathbb{Z}^N \times [0, \infty)$, then the forward equation of $P_{(Y,v)}(x_1, \cdots, x_N, \pi; t)$ for any $x_1 < \cdots < x_N$ is in the form of the (π, v)-element of

$$\frac{d}{dt} \mathbf{U}(x_1, \cdots, x_N; t) = \sum_{j=1}^{N} \mathbf{r}_j \mathbf{U}(x_1, \cdots, x_{j-1}, x_j - 1, x_{j+1}, \cdots, x_N; t) - (\mathbf{r}_1 + \cdots + \mathbf{r}_N) \mathbf{U}(x_1, \cdots, x_N; t) \tag{11}$$

subject to the (π, v)-element of

$$\begin{aligned}
\mathbf{r}_{i+1} \mathbf{U}(x_1, \cdots, x_{i-1}, x_i, x_i, x_{i+2}, \cdots, x_N; t) = \\
(\mathbf{B}_i + \mathbf{r}_i - \mathbf{C}_i) \mathbf{U}(x_1, \cdots, x_{i-1}, x_i, x_i + 1, x_{i+2}, \cdots, x_N; t)
\end{aligned} \tag{12}$$

for all $i = 1, \cdots, N - 1$.

2.2. Solutions of the Forward Equations via Bethe Ansatz

The (π, v)-element of (11) is

$$\frac{d}{dt} U_{\pi v}(x_1, \cdots, x_N; t) =$$

$$\sum_{j=1}^{N} b_{\pi(j)} U_{\pi v}(x_1, \cdots, x_{j-1}, x_j - 1, x_{j+1}, \cdots, x_N; t) - (b_{\pi(1)} + \cdots + b_{\pi(N)}) U_{\pi v}(x_1, \cdots, x_N; t).$$

Assume the separation of variables to write $U_{\pi v}(x_1, \cdots, x_N; t) = U_{\pi v}(x_1, \cdots, x_N) T(t)$. Then, the equation of the spatial variables is

$$\begin{aligned}
\varepsilon U_{\pi v}(x_1, \cdots, x_N) = b_{\pi(1)} U_{\pi v}(x_1 - 1, x_2, \cdots, x_N) + \cdots \\
+ b_{\pi(N)} U_{\pi v}(x_1, \cdots, x_{N-1}, x_N - 1) - (b_{\pi(1)} + \cdots + b_{\pi(N)}) U_{\pi v}(x_1, \cdots, x_N)
\end{aligned} \tag{13}$$

for some constant ε with respect to $t, x_1, \cdots, x_N$. Then, we observe that, for any $\sigma \in S_N$,

$$\prod_{i=1}^{N} \left(b_{\pi(i)} \zeta_{\sigma(i)} \right)^{x_i}, \quad \zeta_1, \cdots, \zeta_N \in \mathbb{C}$$

solves (13) if and only if

$$\varepsilon = \frac{1}{\zeta_1} + \cdots + \frac{1}{\zeta_N} - (b_{\pi(1)} + \cdots + b_{\pi(N)}). \tag{14}$$

Based on the observation in the above, assume that the matrix $\mathbf{U}(x_1, \cdots, x_N; t)$ is invertible and that it is decomposed as

$$\mathbf{U}(x_1, \cdots, x_N; t) = \mathbf{J}(t) \mathbf{U}(x_1, \cdots, x_N)$$

where $\mathbf{J}(t) = [J_{\pi v}(t)]$ is an $N^N \times N^N$ diagonal matrix where $J_{\pi\pi}(t)$ are functions of time only. Hence, from (11), we obtain

$$\begin{aligned}
\mathbf{J}(t)^{-1} \left(\frac{d}{dt} \mathbf{J}(t) \right) = \Big(\mathbf{r}_1 \mathbf{U}(x_1 - 1, x_2, \cdots, x_N) + \cdots + \mathbf{r}_N \mathbf{U}(x_1, \cdots, x_{N-1}, x_N - 1) \\
- (\mathbf{r}_1 + \cdots + \mathbf{r}_N) \mathbf{U}(x_1, \cdots, x_N) \Big) \mathbf{U}(x_1, \cdots, x_N)^{-1}.
\end{aligned} \tag{15}$$

Both sides of (15) must be a diagonal matrix $\mathbf{E} = [\varepsilon_{\pi\pi}]$ whose elements are some constants with respect to $t, x_1, \cdots, x_N$. Thus, we obtain the matrix equation for spatial variables

$$\begin{aligned}
\mathbf{E} \mathbf{U}(x_1, \cdots, x_N) = \mathbf{r}_1 \mathbf{U}(x_1 - 1, x_2, \cdots, x_N) + \cdots + \mathbf{r}_N \mathbf{U}(x_1, \cdots, x_{N-1}, x_N - 1) \\
- (\mathbf{r}_1 + \cdots + \mathbf{r}_N) \mathbf{U}(x_1, \cdots, x_N)
\end{aligned} \tag{16}$$

and the matrix equation for the time variable

$$\frac{d}{dt}\mathbf{J}(t) = \mathbf{J}(t)\mathbf{E} = \mathbf{E}\mathbf{J}(t).$$

Lemma 2. *Let* $\mathbf{D}(x_1,\cdots,x_N) = \left[D_{\pi v}(x_1,\cdots,x_N)\right]$ *be an* $N^N \times N^N$ *diagonal matrix with*

$$D_{\pi\pi}(x_1,\cdots,x_N) = b_{\pi(1)}^{x_1} \cdots b_{\pi(N)}^{x_N}.$$

Then, for any $\sigma \in \mathcal{S}_N$,

$$\mathbf{U}(x_1,\cdots,x_N) = \mathbf{D}(x_1,\cdots,x_N)\mathbf{A}\prod_{i=1}^{N} \zeta_{\sigma(i)}^{x_i} \tag{17}$$

where $\mathbf{A}$ *is an arbitrary invertible* $N^N \times N^N$ *matrix whose elements are constants with respect to* $x_1,\cdots,x_N$ *is a solution of* (16) *if and only if* $\varepsilon_{\pi\pi}$ *is given by*

$$\varepsilon_{\pi\pi} = \frac{1}{\zeta_1} + \cdots + \frac{1}{\zeta_N} - \left(b_{\pi(1)} + \cdots + b_{\pi(N)}\right). \tag{18}$$

Proof. First, we observe that

$$\mathbf{r}_i\mathbf{D}(x_1,\cdots,x_{i-1},x_i - 1,x_{i+1},\cdots,x_N) = \mathbf{D}(x_1,\cdots,x_N)$$

because $\mathbf{r}_i$ is a diagonal matrix whose (π,π)-element is $b_{\pi(i)}$. We observe that

$$\mathbf{D}^{-1}(x_1,\cdots,x_N) = \mathbf{D}(x_1^{-1},\cdots,x_N^{-1}),$$

and

$$\mathbf{D}(x_1,\cdots,x_N)\mathbf{D}(y_1,\cdots,y_N) = \mathbf{D}(x_1 + y_1,\cdots,x_N + y_N).$$

Now, we prove the statement. Suppose that (17) is a solution of (16). Substituting (17) into (16) and then dividing both sides by $\prod_{i=1}^{N} \zeta_{\sigma(i)}^{x_i}$, then

$$\mathbf{E}\mathbf{D}(x_1,\cdots,x_N)\mathbf{A} = \mathbf{r}_1\mathbf{D}(x_1 - 1,x_2,\cdots,x_N)\mathbf{A}\zeta_{\sigma(1)}^{-1} + \cdots$$

$$+ \mathbf{r}_N\mathbf{D}(x_1,\cdots,x_{N-1},x_N - 1)\mathbf{A}\zeta_{\sigma(N)}^{-1} - (\mathbf{r}_1 + \cdots + \mathbf{r}_N)\mathbf{D}(x_1,\cdots,x_N)\mathbf{A}$$

$$= \mathbf{D}(x_1,\cdots,x_N)\mathbf{A}\zeta_{\sigma(1)}^{-1} + \cdots + \mathbf{D}(x_1,\cdots,x_N)\mathbf{A}\zeta_{\sigma(N)}^{-1}$$

$$- (\mathbf{r}_1 + \cdots + \mathbf{r}_N)\mathbf{D}(x_1,\cdots,x_N)\mathbf{A}.$$

Multiplying by $\mathbf{A}^{-1}\mathbf{D}^{-1}(x_1,\cdots,x_N)$ on both sides, we obtain

$$\mathbf{E} = \left(\frac{1}{\zeta_{\sigma(1)}} + \cdots + \frac{1}{\zeta_{\sigma(N)}}\right)\mathbf{I}_{N^N} - (\mathbf{r}_1 + \cdots + \mathbf{r}_N),$$

and thus, the (π,π)-element of $\mathbf{E}$ is given by (18). The second part of the proof can be done via the reverse way of the first part of the proof. $\square$

The previous lemma implies that the general solution of (16) is given by

$$\mathbf{U}(x_1,\cdots,x_N) = \sum_{\sigma \in \mathcal{S}_N} \mathbf{D}(x_1,\cdots,x_N)\mathbf{A}_\sigma \prod_{i=1}^{N} \zeta_{\sigma(i)}^{x_i}. \tag{19}$$

2.3. Boundary Conditions

Now, (19) should satisfy the spatial part of the boundary condition (12), that is,

$$\mathbf{r}_{i+1}\mathbf{U}(x_1,\cdots,x_{i-1},x_i,x_i,x_{i+2},\cdots,x_N) =$$
$$(\mathbf{B}_i + \mathbf{r}_i - \mathbf{C}_i)\mathbf{U}(x_1,\cdots,x_{i-1},x_i,x_i+1,x_{i+2},\cdots,x_N) \tag{20}$$

for $i = 1,\cdots,N-1$. Extending the technique used in [4], we will find the formulas of $\mathbf{A}_\sigma$ in (19), which satisfy (20). Define an $N \times N$ diagonal matrix,

$$\mathbf{r}^{x_i} := \begin{bmatrix} b_1^{x_i} & & \\ & \ddots & \\ & & b_N^{x_i} \end{bmatrix}$$

and recall the definition of $\mathbf{R}_{\beta\alpha}$ in (1). Then, we observe that

$$\mathbf{R}_{\beta\alpha} = -\left[\left(\mathbf{I}_{N^2} - \zeta_\alpha(\mathbf{B} + \mathbf{r}\otimes\mathbf{I}_N - \mathbf{C})\right)\left(\mathbf{r}^{x_i}\otimes\mathbf{r}^{x_i+1}\right)\right]^{-1}\left[\left(\mathbf{I}_{N^2} - \zeta_\beta(\mathbf{B} + \mathbf{r}\otimes\mathbf{I}_N - \mathbf{C})\right)\left(\mathbf{r}^{x_i}\otimes\mathbf{r}^{x_i+1}\right)\right].$$

Lemma 3. *If*

$$\mathbf{A}_{T_i\sigma} = \left(\mathbf{I}_{N^{i-1}} \otimes \mathbf{R}_{\sigma(i+1)\sigma(i)} \otimes \mathbf{I}_{N^{N-i-1}}\right)\mathbf{A}_\sigma \tag{21}$$

for all even permutations σ and $i = 1,\cdots,N-1$, then (20) is satisfied.

Proof. First, we note that

$$\mathbf{r}_{i+1} = \underbrace{\mathbf{I}_N \otimes \cdots \otimes \mathbf{I}_N}_{i-1} \otimes (\mathbf{I}_N \otimes \mathbf{r}) \otimes \underbrace{\mathbf{I}_N \otimes \cdots \otimes \mathbf{I}_N}_{N-i-1},$$
$$\mathbf{B}_i + \mathbf{r}_i - \mathbf{C}_i = \underbrace{\mathbf{I}_N \otimes \cdots \otimes \mathbf{I}_N}_{i-1} \otimes (\mathbf{B} + \mathbf{r}\otimes\mathbf{I}_N - \mathbf{C}) \otimes \underbrace{\mathbf{I}_N \otimes \cdots \otimes \mathbf{I}_N}_{N-i-1},$$

and

$$\mathbf{D}(x_1,\cdots,x_N) = \mathbf{r}^{x_1} \otimes \cdots \otimes \mathbf{r}^{x_N}.$$

Substituting (19) into (20), we obtain

$$\sum_{\sigma\in\mathcal{S}_N}\left[\left(\mathbf{r}^{x_1} \otimes \cdots \otimes \mathbf{r}^{x_{i-1}} \otimes \left(\mathbf{r}^{x_i}\otimes\mathbf{r}^{x_i+1}\right) \otimes \mathbf{r}^{x_{i+2}} \otimes \cdots \otimes \mathbf{r}^{x_N}\right) - \right.$$
$$\left. \left(\mathbf{r}^{x_1} \otimes \cdots \otimes \mathbf{r}^{x_{i-1}} \otimes (\mathbf{B} + \mathbf{r}\otimes\mathbf{I}_N - \mathbf{C})\left(\mathbf{r}^{x_i}\otimes\mathbf{r}^{x_i+1}\right) \otimes \mathbf{r}^{x_{i+2}} \otimes \cdots \otimes \mathbf{r}^{x_N}\right)\zeta_{\sigma(i+1)}\right] \tag{22}$$
$$\times \mathbf{A}_\sigma \zeta_{\sigma(i)}^{x_i}\zeta_{\sigma(i+1)}^{x_i}\prod_{j\neq i,i+1}\zeta_{\sigma(j)}^{x_j} = 0.$$

If we express (22) as a sum over the alternating group $\mathcal{A}_N$

$$\sum_{\sigma\in\mathcal{A}_N}\left(\left[\left(\mathbf{r}^{x_1} \otimes \cdots \otimes \mathbf{r}^{x_{i-1}} \otimes \left(\mathbf{r}^{x_i}\otimes\mathbf{r}^{x_i+1}\right) \otimes \mathbf{r}^{x_{i+2}} \otimes \cdots \otimes \mathbf{r}^{x_N}\right) - \right.\right.$$
$$\left. \left(\mathbf{r}^{x_1} \otimes \cdots \otimes \mathbf{r}^{x_{i-1}} \otimes (\mathbf{B} + \mathbf{r}\otimes\mathbf{I}_N - \mathbf{C})\left(\mathbf{r}^{x_i}\otimes\mathbf{r}^{x_i+1}\right) \otimes \mathbf{r}^{x_{i+2}} \otimes \cdots \otimes \mathbf{r}^{x_N}\right)\zeta_{\sigma(i+1)}\right]\mathbf{A}_\sigma$$
$$+ \left[\left(\mathbf{r}^{x_1} \otimes \cdots \otimes \mathbf{r}^{x_{i-1}} \otimes \left(\mathbf{r}^{x_i}\otimes\mathbf{r}^{x_i+1}\right) \otimes \mathbf{r}^{x_{i+2}} \otimes \cdots \otimes \mathbf{r}^{x_N}\right) - \right. \tag{23}$$
$$\left.\left. \left(\mathbf{r}^{x_1} \otimes \cdots \otimes \mathbf{r}^{x_{i-1}} \otimes (\mathbf{B} + \mathbf{r}\otimes\mathbf{I}_N - \mathbf{C})\left(\mathbf{r}^{x_i}\otimes\mathbf{r}^{x_i+1}\right) \otimes \mathbf{r}^{x_{i+2}} \otimes \cdots \otimes \mathbf{r}^{x_N}\right)\zeta_{\sigma(i)}\right]\mathbf{A}_{T_i\sigma}\right)$$
$$\times \zeta_{\sigma(i)}^{x_i}\zeta_{\sigma(i+1)}^{x_i}\prod_{j\neq i,i+1}\zeta_{\sigma(j)}^{x_j} = 0.$$

However, (23) is satisfied if

$$\left[\left(\mathbf{r}^{x_1} \otimes \cdots \otimes \mathbf{r}^{x_{i-1}} \otimes \left(\mathbf{r}^{x_i} \otimes \mathbf{r}^{x_i+1}\right) \otimes \mathbf{r}^{x_{i+2}} \otimes \cdots \otimes \mathbf{r}^{x_N}\right) - \right.$$
$$\left(\mathbf{r}^{x_1} \otimes \cdots \otimes \mathbf{r}^{x_{i-1}} \otimes \left(\mathbf{B} + \mathbf{r} \otimes \mathbf{I}_N - \mathbf{C}\right)\left(\mathbf{r}^{x_i} \otimes \mathbf{r}^{x_i+1}\right) \otimes \mathbf{r}^{x_{i+2}} \otimes \cdots \otimes \mathbf{r}^{x_N}\right)\xi_{\sigma(i+1)}\Big]\mathbf{A}_\sigma$$
$$= - \left[\left(\mathbf{r}^{x_1} \otimes \cdots \otimes \mathbf{r}^{x_{i-1}} \otimes \left(\mathbf{r}^{x_i} \otimes \mathbf{r}^{x_i+1}\right) \otimes \mathbf{r}^{x_{i+2}} \otimes \cdots \otimes \mathbf{r}^{x_N}\right) - \right.$$
$$\left(\mathbf{r}^{x_1} \otimes \cdots \otimes \mathbf{r}^{x_{i-1}} \otimes \left(\mathbf{B} + \mathbf{r} \otimes \mathbf{I}_N - \mathbf{C}\right)\left(\mathbf{r}^{x_i} \otimes \mathbf{r}^{x_i+1}\right) \otimes \mathbf{r}^{x_{i+2}} \otimes \cdots \otimes \mathbf{r}^{x_N}\right)\xi_{\sigma(i)}\Big]\mathbf{A}_{T_i\sigma}$$

for each even permutation σ, which is equivalent to (21). $\square$

In fact, (3) and (5) implies that the assumption of Lemma 3 is satisfied, hence (19) with $\mathbf{A}_\sigma$ in (5) satisfies (20).

2.4. Consistency Relations—Proof of Lemma 1

Lemma 1 confirms that the multi-species TASEP is *integrable* even when the rates are species-dependent.

2.4.1. Proof of Lemma 1 (a)

It suffices to show that

$$(\mathbf{R}_{\alpha\beta} \otimes \mathbf{I}_N \otimes \mathbf{I}_N)(\mathbf{I}_N \otimes \mathbf{I}_N \otimes \mathbf{R}_{\gamma\delta}) = (\mathbf{I}_N \otimes \mathbf{I}_N \otimes \mathbf{R}_{\gamma\delta})(\mathbf{R}_{\alpha\beta} \otimes \mathbf{I}_N \otimes \mathbf{I}_N).$$

This equality clearly holds because both sides are equal to $\mathbf{R}_{\alpha\beta} \otimes \mathbf{R}_{\gamma\delta}$. $\square$

2.4.2. Proof of Lemma 1 (b)—Yang-Baxter Equation

It suffices to show

$$(\mathbf{R}_{\gamma\beta} \otimes \mathbf{I}_N)(\mathbf{I}_N \otimes \mathbf{R}_{\gamma\alpha})(\mathbf{R}_{\beta\alpha} \otimes \mathbf{I}_N) = (\mathbf{I}_N \otimes \mathbf{R}_{\beta\alpha})(\mathbf{R}_{\gamma\alpha} \otimes \mathbf{I}_N)(\mathbf{I}_N \otimes \mathbf{R}_{\gamma\beta}). \tag{24}$$

If we re-arrange the columns and the rows of the $N^3 \times N^3$ matrices in (24) so that all their labels from the same multi-set $[i, j, k]$ are grouped together, then the matrices in (24) become block-diagonal (See Figure 1).

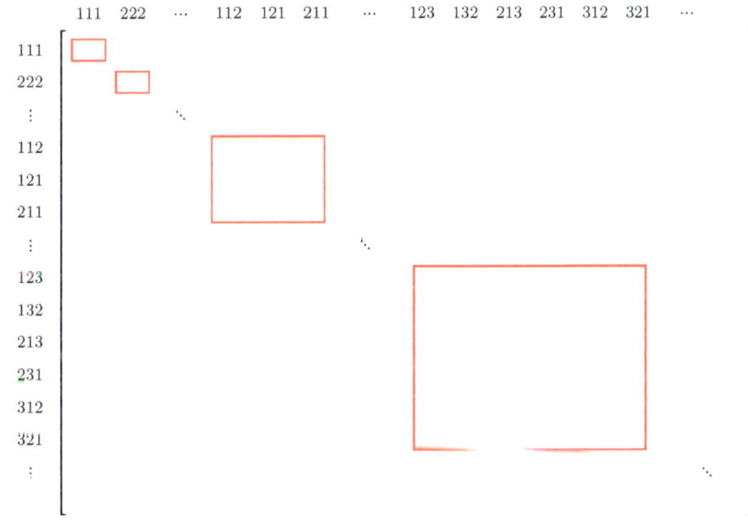

Figure 1. Form of the $N^3 \times N^3$ matrices in (24) after re-ordering the rows and columns.

Let $(\mathbf{R} \otimes \mathbf{I}_N)_{[i,j,k]}$ be the sub-matrix of $(\mathbf{R} \otimes \mathbf{I}_N)$ whose rows and columns are labelled by the permutations of the multi-set $[i, j, k]$, and similarly, we define $(\mathbf{I}_N \otimes \mathbf{R})_{[i,j,k]}$. Then, in order to show (24), it suffices to show

$$(\mathbf{R}_{\gamma\beta} \otimes \mathbf{I}_N)_{[i,j,k]}(\mathbf{I}_N \otimes \mathbf{R}_{\gamma\alpha})_{[i,j,k]}(\mathbf{R}_{\beta\alpha} \otimes \mathbf{I}_N)_{[i,j,k]} =$$
$$(\mathbf{I}_N \otimes \mathbf{R}_{\beta\alpha})_{[i,j,k]}(\mathbf{R}_{\gamma\alpha} \otimes \mathbf{I}_N)_{[i,j,k]}(\mathbf{I}_N \otimes \mathbf{R}_{\gamma\beta})_{[i,j,k]} \tag{25}$$

for each multi-set $[i, j, k]$ whose elements are from $\{1, 2, 3\}$ because all matrices are block-diagonal matrices in the same form. If $i = j = k$, (25) is equivalent to

$$S_{\gamma\beta}(i)S_{\gamma\alpha}(i)S_{\beta\alpha}(i) = S_{\beta\alpha}(i)S_{\gamma\alpha}(i)S_{\gamma\beta}(i),$$

which is trivially true. If $i = j > k$, then (25) is equivalent to

$$\begin{bmatrix} S_{\gamma\beta}(k) & T_{\gamma\beta}(k) & 0 \\ 0 & -1 & 0 \\ 0 & 0 & S_{\gamma\beta}(i) \end{bmatrix} \begin{bmatrix} S_{\gamma\alpha}(i) & 0 & 0 \\ 0 & S_{\gamma\alpha}(k) & T_{\gamma\alpha}(k) \\ 0 & 0 & -1 \end{bmatrix} \begin{bmatrix} S_{\beta\alpha}(k) & T_{\beta\alpha}(k) & 0 \\ 0 & -1 & 0 \\ 0 & 0 & S_{\beta\alpha}(i) \end{bmatrix}$$
$$= \begin{bmatrix} S_{\beta\alpha}(i) & 0 & 0 \\ 0 & S_{\beta\alpha}(k) & T_{\beta\alpha}(k) \\ 0 & 0 & -1 \end{bmatrix} \begin{bmatrix} S_{\gamma\alpha}(k) & T_{\gamma\alpha}(k) & 0 \\ 0 & -1 & 0 \\ 0 & 0 & S_{\gamma\alpha}(i) \end{bmatrix} \begin{bmatrix} S_{\gamma\beta}(i) & 0 & 0 \\ 0 & S_{\gamma\beta}(k) & T_{\gamma\beta}(k) \\ 0 & 0 & -1 \end{bmatrix},$$

which can be easily verified by direct computation. Similarly, the other two cases of (25) for $i = j < k$ and for the case that all i, j, k are distinct can be verified by direct computation. $\square$

2.4.3. Proof of Lemma 1 (c)

It suffices to show that $\mathbf{R}_{\beta\alpha}\mathbf{R}_{\alpha\beta} = \mathbf{I}_{N^2}$. Let us re-arrange the rows and the columns in the same way as in the proof of Lemma 1 (b) to make $\mathbf{R}_{\beta\alpha}$ and $\mathbf{R}_{\alpha\beta}$ block-diagonal. Then, each block on the diagonal of $\mathbf{R}_{\beta\alpha}$ is either a 1×1 matrix or a 2×2 matrix. The 1×1 sub-matrix of $\mathbf{R}_{\beta\alpha}$ consisting of the row ii and the column ii is $[S_{\beta\alpha}(i)]$. The 2×2 sub-matrix of $\mathbf{R}_{\beta\alpha}$ consisting of the rows ij, ji and the columns ij, ji with $i < j$ is

$$\begin{bmatrix} S_{\beta\alpha}(i) & T_{\beta\alpha}(i) \\ 0 & -1 \end{bmatrix}.$$

Similarly, the 1×1 sub-matrix of $\mathbf{R}_{\alpha\beta}$ consisting of the row ii and the column ii is $[S_{\alpha\beta}(i)]$, and the 2×2 sub-matrix of $\mathbf{R}_{\alpha\beta}$ consisting of the rows ij, ji and the columns ij, ji with $i < j$ is

$$\begin{bmatrix} S_{\alpha\beta}(i) & T_{\alpha\beta}(i) \\ 0 & -1 \end{bmatrix}.$$

It is trivial that $S_{\beta\alpha}(i)S_{\alpha\beta}(i) = 1$, and this can be verified,

$$\begin{bmatrix} S_{\beta\alpha}(i) & T_{\beta\alpha}(i) \\ 0 & -1 \end{bmatrix} \begin{bmatrix} S_{\alpha\beta}(i) & T_{\alpha\beta}(i) \\ 0 & -1 \end{bmatrix} = \begin{bmatrix} 1 & 0 \\ 0 & 1 \end{bmatrix},$$

by the direct computation. $\square$

2.5. Initial Condition

The contour integral of (19) multiplied by $\mathbf{J}(t)$ from the left and by $\mathbf{D}'(y_1, \cdots, y_N) \prod_i \zeta_i^{-y_i-1}$ from the right, that is, the right-hand side of (6) still satisfies (11) and (12). (The contour is the one introduced in Theorem 1). Hence, it remains to show that all transition probabilities $P_{(Y,\nu)}(X, \pi; t)$ satisfy the initial condition

$$P_{(Y,\nu)}(X,\pi;0) \;=\; \sum_{\sigma \in \mathcal{S}_N} \oint_c \cdots \oint_c \Lambda_\sigma^{\pi\nu} \prod_{i=1}^{N} b_{\pi(i)}^{x_i} b_{\nu(i)}^{-y_i} \prod_{i=1}^{N} \left(\xi_{\sigma(i)}^{x_i - y_{\sigma(i)} - 1} \right) d\xi_1 \cdots d\xi_N$$

$$= \begin{cases} 1 & \text{if } (Y,\nu) = (X,\pi); \\ 0 & \text{otherwise} \end{cases} \tag{26}$$

where $A_\sigma^{\pi\nu}$ is the (π,ν)-element of $\mathbf{A}_\sigma$. We will show that the integral with the identity permutation in the sum satisfies (26), and other integral terms with non-identity permutations are zero.

Proof. If σ is the identity permutation, then $\mathbf{A}_\sigma$ is the identity matrix. Hence, if $\pi \neq \nu$, then the integral is zero. It is easy to see that if $\pi = \nu$ and $x_i = y_i$ for all i, then the integral is 1. If $\pi = \nu$ and $x_i > y_i$ for some i (recall that our model is totally asymmetric), then the integral becomes zero when integrating with respect to ξ_i. Now, suppose that σ is not the identity permutation. Note that the factors in $A_\sigma^{\pi\nu}$ are from (1), all poles arising from $A_\sigma^{\pi\nu}$, if any, are outside the contours. There exists an i such that $x_i - y_{\sigma(i)} - 1 \geq 0$ because each $x_i \geq y_i$ and σ is not the identity permutation. Hence, integrating with respect to i, the integral is 0. $\square$

3. Conclusions

In this paper, we have shown that the Bethe ansatz method is still applicable to the multi-species TASEP with species-dependent rates. Theorem 1 provides the transition probabilities for all possible compositions of species, which is expected to be used to study further objects, such as the current distribution for certain special initial configurations. The methods used in this paper have limitations in extending to the ASEP ($0 < p < 1$) for now, but it would be interesting to see if the methods can be used to study the species-inhomogeneity of other multi-species models.

Funding: This research was funded by Nazarbayev University (Faculty Development Competitive Grant grant number 090118FD5341 and 021220FD4251.

Institutional Review Board Statement: Not applicable.

Informed Consent Statement: Not applicable.

Data Availability Statement: Not applicable.

Acknowledgments: The author is grateful to anonymous referees for providing valuable comments.

Conflicts of Interest: The author declares no conflict of interest.

References

1. Chatterjee, S.; Schütz, G. Determinant representation for some transition probabilities in the TASEP with second class particles. *J. Stat. Phys.* **2010**, *140*, 900–916. [CrossRef]
2. Kuan, J. Determinantal expressions in multi-species TASEP. *Symmetry Integr. Geom. Methods Appl.* **2020**, *16*, 133.
3. Lee, E. On the TASEP with the second class particles. *Symmetry Integr. Geom. Methods Appl.* **2018**, *14*, 006. [CrossRef]
4. Lee, E. Exact Formulas of the Transition Probabilities of the Multi-Species Asymmetric Simple Exclusion Process. *Symmetry Integr. Geom. Methods Appl.* **2020**, *16*, 139.
5. Tracy, C.; Widom, H. On the asymmetric simple exclusion process with multiple species. *J. Stat. Phys.* **2013**, *150*, 457–470. [CrossRef]
6. Ferrari, P.L.; Ghosal, P.; Nejjar, P. Limit law of a second class particle in TASEP with non-random initial condition. *Ann. Inst. Henri Poincaré Probab. Stat.* **2019**, *55*, 1203–1255. [CrossRef]
7. Nejjar, P. KPZ Statistics of Second Class Particles in ASEP via Mixing. *Commun. Math. Phys.* **2020**, *378*, 601–623. [CrossRef]
8. Borodin, A.; Bufetov, A. Color-position symmetry in interacting particle systems. *Ann. Probab.* **2021**, *49*, 1607–1632. [CrossRef]
9. Borodin, A.; Wheeler, M. Coloured stochastic vertex models and their specctral theory. *arXiv* **2018**, arXiv:1808.01866.
10. Baxter, R.J. *Exactly Solved Models in Statistical Mechanics*; Academic Press: San Diego, CA, USA, 1989.
11. Karbach, M.; Muller, G. Introduction to the Bethe ansatz I. *arXiv* **1998**, arXiv:cond-mat/9809162v1.

12. Rákos, A.; Schütz, G.M. Bethe ansatz and current distribution for the TASEP with particle-dependent hopping rates. *Markov Process. Relat. Fields* **2006**, *12*, 323–334.
13. Lee, E.; Wang, D. Distributions of a particle's position and their asymptotics in the q-deformed totally asymmetric zero range process with site dependent jumping rates. *Stoch. Process. Their Appl.* **2019**, *129*, 1795–1828. [CrossRef]
14. Wang, D.; Waugh, D. The transition probability of the q-TAZRP (q-Bosons) with inhomogeneous jump rates. *Symmetry Integr. Geom. Methods Appl.* **2016**, *12*, 037. [CrossRef]

Article

Shock Structure and Relaxation in the Multi-Component Mixture of Euler Fluids

Damir Madjarević [1,†], Milana Pavić-Čolić [2,†] and Srboljub Simić [2,*,†]

[1] Department of Mechanics, Faculty of Technical Sciences, University of Novi Sad, 21000 Novi Sad, Serbia; damirm@uns.ac.rs

[2] Department of Mathematics and Informatics, Faculty of Sciences, University of Novi Sad, 21000 Novi Sad, Serbia; milana.pavic@dmi.uns.ac.rs

* Correspondence: ssimic@uns.ac.rs

† These authors contributed equally to this work.

Abstract: The shock structure problem is studied for a multi-component mixture of Euler fluids described by the hyperbolic system of balance laws. The model is developed in the framework of extended thermodynamics. Thanks to the equivalence with the kinetic theory approach, phenomenological coefficients are computed from the linearized weak form of the collision operator. Shock structure is analyzed for a three-component mixture of polyatomic gases, and for various combinations of parameters of the model (Mach number, equilibrium concentrations and molecular mass ratios). The analysis revealed that three-component mixtures possess distinguishing features different from the binary ones, and that certain behavior may be attributed to polyatomic structure of the constituents. The multi-temperature model is compared with a single-temperature one, and the difference between the mean temperatures of the mixture are computed. Mechanical and thermal relaxation times are computed along the shock profiles, and revealed that the thermal ones are smaller in the case discussed in this study.

Keywords: shock structure; extended thermodynamics; kinetic theory of gases

Citation: Madjarević, D.; Pavić-Čolić, M.; Simić, S. Shock Structure and Relaxation in the Multi-Component Mixture of Euler Fluids. *Symmetry* **2021**, *13*, 955. https://doi.org/10.3390/sym13060955

Academic Editors: Rahmat Ellahi and Iver H. Brevik

Received: 14 April 2021
Accepted: 19 May 2021
Published: 27 May 2021

Publisher's Note: MDPI stays neutral with regard to jurisdictional claims in published maps and institutional affiliations.

1. Introduction

A shock wave is a paradigm of non-equilibrium process with irreversible character. It drives the system out of equilibrium and, at the same time, causes irreversible changes observed through the increase of entropy [1–3]. In idealized models, shock waves are described as singular surfaces on which jumps of field variables occur. However, when dissipation is taken into account, shock waves are smeared out into continuous profiles, thus creating the so-called shock structure. [4].

Analysis of the shock structure in gaseous media is strongly influenced by the model taken to describe dissipation. For the present study, two approaches are significant—continuum (macroscopic) and kinetic (mesoscopic) one. Continuum approach uses macroscopic field variables and their governing equations to describe the system. Within it, classical continuum models assume non-local constitutive relations for non-convective fluxes like stress tensor, heat flux, or diffusion flux. Typical examples are Navier–Stokes model, Fourier law and Fick law [5]. Such models usually lead to parabolic partial differential equations, or their generalizations. In the same (continuum) setting, dissipation may also be described by means of relaxation, which assumes that non-equilibrium variables converge (relax) towards their equilibrium values during the process. They yield models which usually consist of the hyperbolic systems of balance laws. It is remarkable that in the classical (thermodynamic) limit, i.e., in the small relaxation time approximation, one may recover the classical parabolic models starting from the hyperbolic ones [6–9]. In more physical terms, classical models can be regarded as approximations of the hyperbolic ones when processes occur in the neighborhood of the local equilibrium state [10].

The kinetic approach is focused on the smaller (meso) scale. It takes statistical description—single particle velocity distribution function—and mutual interaction between the particles which causes its change. It is built around the Boltzmann equation and inherits dissipation by construction, through the collision operator [11]. A remarkable feature of the kinetic theory is that it can recover classical continuum models in the so-called hydrodynamic limit, when the Knudsen number is small, using the Chapman-Enskog asymptotic expansion. On the other hand, by means of the moment method kinetic theory yields hyperbolic systems of balance laws for macroscopic observables, taken as moments of the velocity distribution function. Under appropriate assumptions, relaxation continuum models of rarefied gases are completely equivalent to moment equations of the kinetic theory of gases [12,13].

Either approach described above has its advantages and shortcomings. For example, kinetic theory is obviously more refined than the continuum one, but evolution of the system is eventually monitored through macroscopic observables. In this study we shall try to take the benefits from both approaches and analyze the shock structure in the hyperbolic model of the multi-component mixture of Euler fluids. By an Euler fluid we consider the fluid in which viscous and heat conducting effects are negligible. By a phrase multi-component mixture, although it seems to be a pleonasm, we reveal our intention to analyze the mixture of more than two constituents (as opposed to binary, i.e., two-component mixture). In fact, we shall analyze the shock structure in a three-component mixture as a paradigmatic example of the multi-component one. It is known that a multi-component mixture exhibits peculiar phenomena with respect to the binary one, as for instance the uphill diffusion [14–16]. Before we embark on the shock structure analysis, a short review of modelling issues and important results in this particular problem relevant to our study will be given.

Our analysis will be based upon the continuum model of mixture developed within the framework of rational extended thermodynamics (RET) [13,17] (for a review of the model one may consult [18]). This model is rather peculiar from macroscopic point of view since it takes velocities and temperatures of the constituents as field variables. RET formalism, i.e., Galilean invariance of the governing equations and their compatibility with the entropy inequality, makes the model thermodynamically consistent. The model is hyperbolic, and dissipation is taken into account through mutual interaction between the constituents, which turns out to be of the relaxation type. However, this approach is limited in the sense that phenomenological coefficients, which appear in the closure process, cannot be explicitly determined within its framework. To resolve this problem, we turn to the kinetic approach based upon the polyatomic gas mixture model, i.e., the system of Boltzmann equations, with the continuous internal energy variable [19–21], which is more tractable than the semi-classical model with discrete energy levels [22–25], yet shown to reproduce physical requirements [26–29] and to be well posed mathematically in the case of space homogeneity [30]. Moreover, the multi-temperature (MT) model for mixture of Euler fluids is fully compatible with moment equations derived from this system at an Euler level [31,32]. In particular, we find the complete closure of the MT model by using the phenomenological coefficients from the weak form of the continuous collision operator at an Euler level [33].

In this study, the shock wave will be treated as a continuous traveling wave solution of the governing equations which converges to equilibrium states at infinity. Such an assumption makes feasible the use of dynamical systems approach. This kind of theoretical shock structure analysis in continuum models of single-component fluids has its roots in [34], where it was applied to Navier–Stokes–Fourier model. Further application of this method to hyperbolic continuum models was studied in [35] (see also Chapter 12 in [13] for a comprehensive review). The main feature of this approach is that the shock structure is described by a heteroclinic orbit in the state space which connects stationary points related to upstream and downstream equilibrium. More precisely, this orbit is an intersection of unstable and stable manifolds of these stationary points [36], and can be related to the bifurcation of the upstream equilibrium point [37]. This could also be useful for the numerical computation of the shock profiles, albeit in the limited way. Namely, it was shown [38] that in hyperbolic systems of balance laws, endowed with a convex entropy density, continuous shock structure does not exist if the shock speed is greater than the highest characteristic speed of the system; instead, there appears a continuous solution with the so-called sub-shock.

Apart from single-component models of gases, shock waves were also studied in the context of mixtures. In the early stage, they were analyzed by means of classical models [39] and DSMC [40]. Efficient computing devices moved the focus towards development of numerical methods for the solution of Boltzmann equations for mixtures [41–43]. On the other hand, shock structure was also analyzed for the moment equations of mixtures, derived from the kinetic theory for various types of interaction [44,45]. At the same time, detailed and systematic analysis of the shock structure in hyperbolic MT mixture of Euler fluids was given in [46,47], which revealed certain new features of the temperature overshoot. In this context, existence of continuous profiles and appearance of sub-shocks turned out to be more delicate than in the one-component case, as discussed in [48–50]. Shock structure in the mixture of noble gases was revisited by DSMC in [51]. Recently, attention was given to the entropy growth and the entropy production analysis in the mixture of Euler fluids [52], and the mixture with viscous and thermal dissipation [53]. It was found out that in former case diffusion makes the major contribution to dissipation, while in the latter one it is the thermal conduction.

It is remarkable that most of the studies mentioned above were concerned with binary mixtures. They certainly reveal the properties which distinguish them from single-component fluids, like temperature overshoot or influence of diffusion on the shock thickness. However, there are phenomena which can be observed only when the multi-component mixture is studied. This gives us strong motivation to analyze the shock structure problem in three-component mixture of Euler fluids. So far, only few references were found which analyze the shock waves in multi-component mixtures within the frameworks indicated above. In [54], the shock structure was analyzed in a mixture of gases with disparate molecular masses, and viscous and thermal dissipation; ternary mixture is studied as a final example, with a special attention on the presence of two zones within the shock profile. In [55], the analysis was based upon numerical solution of the Boltzmann equations; although the main focus was on binary mixture, three-species mixture was studied for the computation of parallel temperatures which exhibit an overshoot. Finally, in [56], three- and four-component mixtures are studied by numerical solution of the Boltzmann equations; profiles of macroscopic variables are found to have good agreement with experiments and computations by other authors.

Our study will be focused on the shock waves in mixtures of Euler fluids, using the MT model developed in [17], together with phenomenological coefficients from [33]. It will be assumed that the mixture is homogeneous in the sense that in each infinitesimal volume element all the constituents are present. It will also be assumed that all the constituents are ideal gases, described by usual thermal and caloric equations of state, and that chemical reactions do not occur between them. Once equipped with the closed, thermodynamically consistent system of governing equations, we shall analyze numerically computed shock structures in three-component mixtures. Our aim is to reveal and bring to light possible qualitative novelties which appear when the mixture contains more than two components.

In the rest of the paper, model of the mixture of Euler fluids is presented in Section 2. It brings together continuum modeling and kinetic theory results to get the closed system of equations. In Section 3, the shock structure problem for a three-component mixture is formulated. It contains all the information needed for derivation of equations, their transformation into dimensionless form, and building up an appropriate numerical procedure. Section 4 contains numerical solutions of the shock structure equations and their analysis. Shock profiles are analyzed for different values of parameters of the model (Mach number, equilibrium concentrations and molecular mass ratios), which provides an insight of their importance in the shock structure analysis. Additionally, multi-temperature shock profiles are compared with single-temperature ones, and the norm of their difference is computed. Finally, Section 5 brings the analysis of relaxation processes captured by the model. It is of utmost importance since they manifest dissipation. Therefore, relative magnitudes of relaxation times for mechanical and thermal relaxations are computed throughout the profile. Paper ends up with proper conclusions and an outlook of future research directions.

2. Multi-Component Mixture of Euler Fluids

A mixture is a medium consisted of more than one identifiable constituent. One of the most comprehensive mathematical models for multi-component mixtures of Euler fluids is developed within the framework of rational extended thermodynamics (RET) [13]. It inherits the postulates, so-called metaphysical principles, which are the basis for rational thermodynamics of mixtures [57], but also exploits the principles of RET—local dependence of constitutive functions on field variables, Galilean invariance of governing equations and entropy principle with the convex entropy density. As a result, governing equations have the structure of hyperbolic system of balance laws. For a complete resolution of the closure problem, it is important to note that these models are fully compatible with transfer equations for moments, obtained from Boltzmann equation(s) by means of the moment method.

2.1. General Structure of the Model

In the RET approach to the mixture modelling it is assumed that the state of each constituent $\alpha = 1, \ldots, n$ is described by its own set of field variables—mass density ρ_α, velocity $\mathbf{v}_\alpha$ and temperature T_α. There lies the main difference with respect to other, more traditional approaches—it is (primarily) multi-velocity, and multi-temperature model. Further assumptions constitute the set of postulates known as metaphysical principles [57]: (i) all properties of the mixture are mathematical consequences of properties of the constituents; (ii) behavior of each constituent is governed by the same equations as if it were isolated from the mixture, but mutual interactions with other constituents must be taken into account; (iii) behavior of the whole mixture is governed by the same equations as is a single body.

Application of postulate (ii) yields governing equations for the constituents:

$$\frac{\partial \rho_\alpha}{\partial t} + \mathrm{div}(\rho_\alpha \mathbf{v}_\alpha) = \tau_\alpha,$$

$$\frac{\partial}{\partial t}(\rho_\alpha \mathbf{v}_\alpha) + \mathrm{div}(\rho_\alpha \mathbf{v}_\alpha \otimes \mathbf{v}_\alpha - \mathbf{t}_\alpha) = \mathbf{m}_\alpha, \tag{1}$$

$$\frac{\partial}{\partial t}\left(\frac{1}{2}\rho_\alpha v_\alpha^2 + \rho_\alpha \varepsilon_\alpha\right) + \mathrm{div}\left\{\left(\frac{1}{2}\rho_\alpha v_\alpha^2 + \rho_\alpha \varepsilon_\alpha\right)\mathbf{v}_\alpha - \mathbf{t}_\alpha \mathbf{v}_\alpha + \mathbf{q}_\alpha\right\} = e_\alpha,$$

where ε_α are the specific internal energies, $\mathbf{t}_\alpha$ are the stress tensors, $\mathbf{q}_\alpha$ are the heat fluxes and τ_α, $\mathbf{m}_\alpha$ and e_α are the source terms describing interaction between the constituents. To apply postulate (iii) we must introduce restriction (axiom) regarding the source terms:

$$\sum_{\alpha=1}^{n} \tau_\alpha = 0, \quad \sum_{\alpha=1}^{n} \mathbf{m}_\alpha = \mathbf{0}, \quad \sum_{\alpha=1}^{n} e_\alpha = 0. \tag{2}$$

Then, summation of Equation (1) yields the conservation laws for the mixture:

$$\frac{\partial \rho}{\partial t} + \mathrm{div}(\rho \mathbf{v}) = 0,$$

$$\frac{\partial}{\partial t}(\rho \mathbf{v}) + \mathrm{div}(\rho \mathbf{v} \otimes \mathbf{v} - \mathbf{t}) = \mathbf{0}, \tag{3}$$

$$\frac{\partial}{\partial t}\left(\frac{1}{2}\rho v^2 + \rho \varepsilon\right) + \mathrm{div}\left\{\left(\frac{1}{2}\rho v^2 + \rho \varepsilon\right)\mathbf{v} - \mathbf{t}\mathbf{v} + \mathbf{q}\right\} = 0,$$

provided we define the mixture variables as follows:

$$\rho = \sum_{\alpha=1}^{n} \rho_\alpha, \quad \mathbf{v} = \frac{1}{\rho} \sum_{\alpha=1}^{n} \rho_\alpha \mathbf{v}_\alpha, \quad \mathbf{u}_\alpha = \mathbf{v}_\alpha - \mathbf{v},$$

$$\varepsilon_I = \frac{1}{\rho} \sum_{\alpha=1}^{n} \rho_\alpha \varepsilon_\alpha, \quad \varepsilon = \varepsilon_I + \frac{1}{2\rho} \sum_{\alpha=1}^{n} \rho_\alpha u_\alpha^2, \tag{4}$$

$$\mathbf{t} = \sum_{\alpha=1}^{n} (\mathbf{t}_\alpha - \rho_\alpha \mathbf{u}_\alpha \otimes \mathbf{u}_\alpha), \quad \mathbf{q} = \sum_{\alpha=1}^{n} \left\{ \mathbf{q}_\alpha + \rho_\alpha \left(\varepsilon_\alpha + \frac{1}{2} u_\alpha^2 \right) \mathbf{u}_\alpha - \mathbf{t}_\alpha \mathbf{u}_\alpha \right\},$$

where ρ is the mixture mass density, $\mathbf{v}$ is the mean velocity, $\mathbf{u}_\alpha$ is the diffusion velocity, ε_I is the intrinsic specific internal energy, ε is the specific internal energy, $\mathbf{t}$ is the mixture stress tensor, and $\mathbf{q}$ is the mixture flux of internal energy. Definitions (4) comprise the postulate (i).

In the sequel we shall introduce further restrictions into our model. First, we shall assume that all the constituents are Euler fluids, i.e., their stress tensors are symmetric and heat fluxes vanish:

$$\mathbf{t}_\alpha = -p_\alpha \mathbf{I}, \quad \mathbf{q}_\alpha = \mathbf{0}, \tag{5}$$

where p_α are partial pressures, and $\mathbf{I}$ is the identity tensor. Second, we shall assume that each constituent obey thermal and caloric equation of state of ideal gas:

$$p_\alpha = \rho_\alpha \frac{k_B}{m_\alpha} T_\alpha, \quad \varepsilon_\alpha = \frac{k_B}{m_\alpha(\gamma_\alpha - 1)} T_\alpha = c_{V\alpha} T_\alpha, \tag{6}$$

where k_B is the Boltzmann constant, and m_α are the molecular masses of the constituents. Ratios of specific heats γ_α and constant volume specific heats $c_{V\alpha}$ of the constituents are assumed to be constant. Finally, we shall assume that there are no mutual chemical interactions between the constituents, which amounts to:

$$\tau_\alpha = 0, \quad \alpha = 1, \ldots, n. \tag{7}$$

To describe behavior of the mixture, $5n$ scalar field variables are used. Thus, $5n$ scalar governing equations are needed, and this choice may not be unique. Here, we follow the standard procedure and choose conservation laws (3) for the mixture, and $n-1$ balance laws (1) (we drop the balance laws for the constituent $\alpha = n$, and use index $b = 1, \ldots, n-1$ instead of α). Such a model is usually called the multi-temperature (MT) model of mixture.

The RET approach is peculiar in the sense that it requires Galilean invariance of governing equations from the outset, and the compatibility with the entropy inequality with convex entropy density. This resolves the closure problem to an extent standard for macroscopic theories. To summarize the consequences of these principles, primarily reflected on the source terms, we emphasize that Galilean invariance restricts their velocity dependence:

$$\mathbf{m}_b = \hat{\mathbf{m}}_b; \quad e_b = \hat{e}_b + \hat{\mathbf{m}}_b \cdot \mathbf{v}, \quad b = 1, \ldots, n-1, \tag{8}$$

while the entropy principle yields more specific structure of velocity independent terms $\hat{\mathbf{m}}_b$ and $\hat{e}_b$:

$$\hat{\mathbf{m}}_b = -\sum_{c=1}^{n-1} \psi_{bc}(\mathbf{w}) \left(\frac{\mathbf{u}_c}{T_c} - \frac{\mathbf{u}_n}{T_n} \right); \quad \hat{e}_b = -\sum_{c=1}^{n-1} \theta_{bc}(\mathbf{w}) \left(-\frac{1}{T_c} + \frac{1}{T_n} \right), \tag{9}$$

where $\psi_{bc}(\mathbf{w})$ and $\theta_{bc}(\mathbf{w})$ are positive semi definite matrix functions of objective quantities $\mathbf{w}$ [17,18].

If the conditions of thermodynamic process are such that large discrepancies between species temperatures are not expected, one may use the single-temperature (ST), but still multi-velocity model. All the constituents have the same temperature T, and governing equations consist of the conservation laws (3) and the balance laws (1)$_{1,2}$, i.e., energy balance laws for the constituents are discarded. This system is a principal subsystem in the hierarchy of hyperbolic systems, as described in [17,58].

Local equilibrium conditions $\mathbf{m}_b = \mathbf{0}$, $e_b = 0$ imply the following conditions:

$$T_1 = T_2 = \ldots = T_n = T;$$
$$\mathbf{u}_1 = \mathbf{u}_2 = \ldots = \mathbf{u}_n(= \mathbf{0}) \quad \Leftrightarrow \quad \mathbf{v}_1 = \mathbf{v}_2 = \ldots = \mathbf{v}_n = \mathbf{v}, \tag{10}$$

i.e., all the constituents have common temperature, and common velocity. Conditions (10) determine the so-called equilibrium subsystem [58].

A brief comment on source terms may shed a light on their role in the model. They are introduced in accordance with metaphysical principles [57]—principle (ii) in particular—and they describe mutual interaction between the constituents. Their form (9) comes out as a consequence of the entropy inequality, and it secures the proper sign of the entropy production rate. In that sense, although the constituents are ideal gases, dissipation is brought into system through the source terms. Finally, their structure reveals also the physical basis of dissipation—it is a consequence of mutual exchange of momentum and energy between the constituents.

2.2. Equations of State and Average Quantities for the Mixture

The use of conservation laws (3) as governing equations implicitly introduces assumption that total pressure p of the mixture and intrinsic specific internal energy ε_I mimic the equations of state of constituents. Namely:

$$p = \rho \frac{k_B}{m} T, \quad \varepsilon_I = \frac{k_B}{(\gamma - 1)m} T. \tag{11}$$

where T is the average temperature of the mixture, m is the average atomic mass, and γ is the average ratio of specific heats. Total pressure and intrinsic specific internal energy must obey Dalton's law and defining relation (4):

$$p = \sum_{\alpha=1}^{n} p_\alpha, \quad \rho \varepsilon_I = \sum_{\alpha=1}^{n} \rho_\alpha \varepsilon_\alpha. \tag{12}$$

In the sequel, we shall introduce the mass concentration variables c_α along with restriction which they obey due to (4):

$$c_\alpha = \frac{\rho_\alpha}{\rho}, \quad \sum_{\alpha=1}^{n} c_\alpha = 1. \tag{13}$$

Since relations (11) and (12) must hold in local equilibrium, $T_1 = \ldots = T_n = T$, the following implicit definitions emerge:

$$\frac{1}{m} = \sum_{\alpha=1}^{n} \frac{c_\alpha}{m_\alpha}, \quad \frac{1}{(\gamma - 1)m} = \sum_{\alpha=1}^{n} \frac{c_\alpha}{(\gamma_\alpha - 1)m_\alpha}, \tag{14}$$

where, obviously, $m = m(c_\alpha)$ and $\gamma = \gamma(c_\alpha)$. On the other hand, in general case, from (11), (12) and (14) one obtains definition of the average temperature T:

$$T = (\gamma - 1)m \sum_{\alpha=1}^{n} \frac{c_\alpha}{(\gamma_\alpha - 1)m_\alpha} T_\alpha. \tag{15}$$

Note that although mixture equations of state (11) resemble the form of the same equations in the single-component gas, they depend on concentrations of the constituents due to (14). Furthermore, the mixture cannot be treated as ideal gas since the stress tensor $\mathbf{t}$, apart from total pressure p, has non-vanishing diagonal and off-diagonal terms, and the flux of internal energy $\mathbf{q}$ does not vanish (see Equation (4)).

2.3. Diffusion Fluxes and Diffusion Temperatures

It is common in the mixture theory to introduce quantities which describe the difference between the state of constituents and certain properly defined state of the mixture. One such example is the diffusion velocity $\mathbf{u}_\alpha = \mathbf{v}_\alpha - \mathbf{v}$. In this study we shall choose the so-called diffusion quantities which frequently appear in the literature. These are the diffusion fluxes $\mathbf{J}_\alpha$ and the diffusion temperatures Θ_α:

$$\mathbf{J}_\alpha = \rho_\alpha \mathbf{u}_\alpha, \quad \Theta_\alpha = T_\alpha - T. \tag{16}$$

Due to (4), (14) and (15), they are not independent and must satisfy the following constraints:

$$\sum_{\alpha=1}^{n} \mathbf{J}_\alpha = \mathbf{0}, \quad \sum_{\alpha=1}^{n} \frac{c_\alpha}{(\gamma_\alpha - 1)m_\alpha} \Theta_\alpha = 0. \tag{17}$$

Therefore, in the sequel whenever appears summation over α, or quantities related to nth constituent appear, the following relations will be assumed:

$$c_n = 1 - \sum_{b=1}^{n-1} c_b; \quad \mathbf{J}_n = -\sum_{b=1}^{n-1} \mathbf{J}_b; \quad \Theta_n = -\sum_{b=1}^{n-1} Y_{bn}\Theta_b, \tag{18}$$

where:

$$Y_{bn} = \frac{c_b}{c_n} \frac{(\gamma_n - 1)m_n}{(\gamma_b - 1)m_b}. \tag{19}$$

Additionally, in definitions of all the mixture variables (4) partial mass densities ρ_α will be replaced with mass concentrations, i.e., $\rho_\alpha = \rho c_\alpha$, and diffusion velocities $\mathbf{u}_\alpha$ will be replaced with diffusion fluxes, i.e., $\mathbf{u}_\alpha = \mathbf{J}_\alpha / \rho_\alpha$.

2.4. Phenomenological Coefficients

As a continuum theory, RET is limited in the sense that phenomenological coefficients, which appear in the source terms (9), cannot be determined explicitly within its framework. This could be overcome by matching with an experimental evidence, or by means of some more refined approach. In this study, the latter path is taken, so that the kinetic theory of gases and moment equations appear as a natural choice. This procedure is completely developed in [33], albeit with slightly different notation. In the sequel, we shall briefly outline these results.

Basis for computation of the phenomenological coefficients, proposed in [33], is the multi-velocity and multi-temperature model of mixture of gases described by means of Boltzmann equations. The model is analyzed at an Euler level, i.e., using Maxwellian velocity distribution function for each constituent, with its own velocity and its own temperature. Moment equations consisted of mass, momentum and energy balance laws. However, source terms obtained as moments of collision operator had a structure that cannot be correlated to (9). Therefore, both source terms were linearized in the neighborhood of the local equilibrium state, and the explicit form of phenomenological coefficients ψ_{bc} and θ_{bc} was determined.

The source terms (9) can be linearized near the local equilibrium state determined with conditions (10). Firstly, the objective quantity $\mathbf{w} - \mathbf{w}(\mathbf{u}_1, \ldots, \mathbf{u}_n, T_1, \ldots, T_n)$ evaluated at the local equilibrium state $(\mathbf{0}, \ldots, \mathbf{0}, T, \ldots, T)$ is denoted with:

$$\mathbf{w}^0 = \mathbf{w}(\mathbf{0}, \ldots, \mathbf{0}, T, \ldots, T). \tag{20}$$

Then the source terms (9) can be approximated by:

$$\hat{\mathbf{m}}_b \approx -\sum_{c=1}^{n-1} \psi_{bc}(\mathbf{w}^0)\left(\frac{\mathbf{u}_c - \mathbf{u}_n}{T}\right); \quad \hat{e}_b \approx -\sum_{c=1}^{n-1} \theta_{bc}(\mathbf{w}^0)\left(\frac{T_c - T_n}{T^2}\right),$$

Such an approximation enables comparison with the source terms computed from the Boltzmann collision operator in a suitable asymptotics. Namely, in [33] the phenomenological coefficients are explicitly computed, depending on the parameter $s_{\alpha\beta}$ of the cross-section, and satisfying $s_{\alpha\beta} = s_{\beta\alpha}$ and $s_{\alpha\beta} > -3/4$ for each $\alpha, \beta = 1, \ldots, n$. In our notation, they read:

$$\psi_{bc}(\mathbf{w}^0) = \begin{cases} -\dfrac{2}{3} T\, \mu_{bc}\, K^\psi_{bc}, & b \neq c, \\[2ex] \dfrac{2}{3} T \displaystyle\sum_{\substack{\ell=1 \\ \ell \neq b}}^{n} \mu_{b\ell} K^\psi_{b\ell}, & b = c, \end{cases} \qquad \theta_{bc}(\mathbf{w}^0) = \begin{cases} -k_B T^2 K^\theta_{bc}, & b \neq c, \\[2ex] k_B T^2 \displaystyle\sum_{\substack{\ell=1 \\ \ell \neq b}}^{n} K^\theta_{b\ell}, & b = c, \end{cases} \tag{21}$$

for any $b, c = 1, \ldots, n-1$. Coefficients $K^\psi_{\alpha\beta}$ and $K^\theta_{\alpha\beta}$ are defined for any $\alpha, \beta = 1, \ldots, n$, as long as $\alpha \neq \beta$, and read as follows:

$$K^\psi_{\alpha\beta} = \mathbb{K}\, \kappa^\psi_{\alpha\beta} \frac{\rho_\alpha}{m_\alpha} \frac{\rho_\beta}{m_\beta} \mu_{\alpha\beta}^{-\frac{s_{\alpha\beta}}{2}} (k_B T)^{-(d_\alpha + d_\beta) + \frac{s_{\alpha\beta}}{2}}, \qquad K^\theta_{\alpha\beta} = \kappa^\theta_{\alpha\beta} K^\psi_{\alpha\beta} \tag{22}$$

where the following constants are introduced:

$$\begin{aligned} \kappa^\psi_{\alpha\beta} &= \frac{2^{\frac{s_{\alpha\beta}}{2}+4}\Gamma\left[\frac{s_{\alpha\beta}+3}{2}\right]}{s_{\alpha\beta}+5} \frac{\sqrt{\pi}}{\Gamma[d_\alpha+1]\Gamma[d_\beta+1]}, \\[2ex] \kappa^\theta_{\alpha\beta} &= \frac{2(s_{\alpha\beta}+5)}{s_{\alpha\beta}+7}\left(\frac{1}{s_{\alpha\beta}+3} + \frac{\mu_{\alpha\beta}}{m_\alpha+m_\beta}\right). \end{aligned} \tag{23}$$

In (22) and (23), $\mu_{\alpha\beta}$ and d_α are the reduced mass and the parameter related to the ratio of specific heats, respectively:

$$\mu_{\alpha\beta} = \frac{m_\alpha m_\beta}{m_\alpha + m_\beta}, \qquad d_\alpha = \frac{-5\gamma_\alpha + 7}{2(\gamma_\alpha - 1)}.$$

Dimensional constant $\mathbb{K}$ in (22) secures the proper dimension of the coefficients, and can be represented as a product of constituent-related part $\mathbb{K}_{\alpha\beta}$ and common term $\mathbb{K}_*$ in the following way:

$$\mathbb{K} = \mathbb{K}_* \,\mathbb{K}_{\alpha\beta}, \qquad \mathbb{K}_{\alpha\beta} = \frac{m_0^2}{\rho_0^2}(k_B T_0)^{d_\alpha + d_\beta}\left(\frac{k_B}{\mu_{\alpha\beta}} T_0\right)^{-\frac{s_{\alpha\beta}}{2}}, \tag{24}$$

where subscript 0 indicates the values of average quantities (mass, density and temperature) in a reference state.

2.5. Relaxation Times

Source terms (9) describe mutual interaction of the constituents and model dissipation in the system through relaxation towards (local) equilibrium. This process is usually expressed in terms of relaxation times—quantities which provide estimates of time-rate of convergence towards equilibrium. They can be related to phenomenological coefficients, and in the case of n-component mixture they read:

$$\tau_{Dbc} = (-1)^{1-\delta_{bc}} \frac{1}{\psi_{bc}(\mathbf{w}^0)} \frac{\rho_c \rho_n}{\sum_{\alpha=1}^{n} \rho_\alpha} T, \qquad \tau_{Tbc} = (-1)^{1-\delta_{bc}} \frac{1}{\theta_{bc}(\mathbf{w}^0)} \frac{\rho_c c_{Vc} \rho_n c_{Vn}}{\sum_{\alpha=1}^{n} \rho_\alpha c_{V\alpha}} T^2, \tag{25}$$

where τ_{Dbc} and τ_{Tbc} denote mechanical (diffusion) and thermal relaxation time, respectively, and factor $(-1)^{1-\delta_{bc}}$, with δ_{bc} being Kronecker delta, is used to secure their positivity. Using relaxation times (25), source terms may be formally expressed in the following form:

$$
\begin{aligned}
\hat{\mathbf{m}}_b &= -\sum_{c=1}^{n-1} (-1)^{1-\delta_{bc}} \frac{1}{\tau_{Dbc}(\mathbf{w}^0)} \frac{\rho_c \rho_n}{\sum_{\alpha=1}^{n} \rho_\alpha} T\left(\frac{\mathbf{u}_c}{T_c} - \frac{\mathbf{u}_n}{T_n} \right); \\
\hat{e}_b &= -\sum_{c=1}^{n-1} (-1)^{1-\delta_{bc}} \frac{1}{\tau_{Tbc}(\mathbf{w}^0)} \frac{\rho_c c_{Vc} \rho_n c_{Vn}}{\sum_{\alpha=1}^{n} \rho_\alpha c_{V\alpha}} T^2\left(-\frac{1}{T_c} + \frac{1}{T_n} \right),
\end{aligned}
\tag{26}
$$

It may be observed that relaxation times (25) present proper generalization of the relaxation times given in [47].

3. Shock Structure Problem for Three-Component Mixture of Euler Fluids

Our main concern in this study is the shock structure problem for three-component mixture of Euler fluids. In idealized situation (non-dissipative models) shock waves are (singular) surfaces, which move through space, on which jumps of field variables occur. If s denotes the shock speed—propagation velocity component perpendicular to shock surface—then the values of field variables in front and behind the shock are related to s by celebrated Rankine–Hugoniot conditions. In our analysis we shall analyze single shock whose surface is plane, and the mixture flows in direction perpendicular to the shock wave. If the dissipation is present in the model, shock wave is smeared out into the shock structure (profile). This is smooth solution, in the form of traveling wave, which has steep gradients in the small neighborhood of the shock wave and asymptotically connects equilibrium states in front (upstream) and behind (downstream) the shock. Equilibrium states are related to the shock speed s through the Rankine–Hugoniot equations.

With these preliminaries at hand, we may sketch the steps which will lead us to a formulation of the shock structure problem for three-component mixture of Euler fluids ($n = 3$). Since the analysis is restricted to plane waves, we shall first expose governing equations for one-dimensional flow. Then, taking the traveling wave solution ansatz we shall derive equations which determine the shock structure. Finally, the shock structure problem will be put into dimensionless form which will be subsequently numerically integrated and analyzed.

3.1. Equations for One-Dimensional Flow

In one dimensional settings, certain simplifications have to be done. If $\mathbf{e}_i$, $i = 1, 2, 3$, represents standard basis in $\mathbb{R}^3$, then we assume motion in $\mathbf{e}_1$ direction and respective coordinate denote by x. Velocities will be $\mathbf{v} = v\mathbf{e}_1$ and $\mathbf{u}_\alpha = u_\alpha \mathbf{e}_1$, and diffusion fluxes $\mathbf{J}_\alpha = J_\alpha \mathbf{e}_1$. Mixture stress tensor will have the form $\mathbf{t} = t_{11}\mathbf{e}_1 \otimes \mathbf{e}_1$, and flux of internal energy will be $\mathbf{q} = q\mathbf{e}_1$, where:

$$
t_{11} = -p - \sum_{\alpha=1}^{3} \frac{1}{\rho_\alpha} J_\alpha^2, \qquad q = \sum_{\alpha=1}^{3} \left(\varepsilon_\alpha + \frac{p_\alpha}{\rho_\alpha} + \frac{1}{2\rho_\alpha^2} J_\alpha^2 \right) J_\alpha,
\tag{27}
$$

whose more explicit form will be given in the sequel. Source terms then read $\hat{\mathbf{m}}_b = \hat{m}_b \mathbf{e}_1$.

Simplifications introduced above lead us to the following one dimensional governing equations, conservation laws for the mixture:

$$
\begin{aligned}
&\frac{\partial \rho}{\partial t} + \frac{\partial}{\partial x}(\rho v) = 0, \\
&\frac{\partial}{\partial t}(\rho v) + \frac{\partial}{\partial x}\left(\rho v^2 - t_{11} \right) = 0, \\
&\frac{\partial}{\partial t}\left(\frac{1}{2}\rho v^2 + \rho \varepsilon \right) + \frac{\partial}{\partial x}\left[\left(\frac{1}{2}\rho v^2 + \rho \varepsilon \right) v - t_{11} v + q \right] = 0,
\end{aligned}
\tag{28}
$$

and balance laws for the constituents:

$$\frac{\partial}{\partial t}(\rho c_b) + \frac{\partial}{\partial x}(\rho c_b v + J_b) = 0,$$

$$\frac{\partial}{\partial t}(\rho c_b v + J_b) + \frac{\partial}{\partial x}\left(\rho c_b v^2 + \frac{J_b^2}{\rho c_b} + 2v J_b + p_b\right) = \hat{m}_b,$$

$$\frac{\partial}{\partial t}\left[\frac{1}{2}\rho c_b\left(v + \frac{J_b}{\rho c_b}\right)^2 + \rho c_b \varepsilon_b\right]$$

$$+ \frac{\partial}{\partial x}\left\{\left[\frac{1}{2}\rho c_b\left(v + \frac{J_b}{\rho c_b}\right)^2 + \rho c_b \varepsilon_b + p_b\right]\left(v + \frac{J_b}{\rho c_b}\right)\right\} = \hat{e}_b + \hat{m}_b v,$$

(29)

where $b = 1, 2$.

3.2. Shock Structure Equations

To derive equations which determine the shock structure we assume the traveling wave profile of the solution. In that case generic field variable $U(t, x)$ depends on a single composite variable $\xi = x - st$, $U = U(\xi)$, where s is the shock speed. Partial derivatives are replaced with ordinary derivatives in the following manner, $\partial/\partial t = -s(d/d\xi)$, $\partial/\partial x = d/d\xi$, and relative velocity with respect to the shock front is defined as $u = v - s$. After straightforward computations our system of governing equations is transformed into ODE system of shock structure equations, i.e., conservation laws for the mixture:

$$\frac{d}{d\xi}(\rho u) = 0,$$

$$\frac{d}{d\xi}\left(\rho u^2 - t_{11}\right) = 0,$$

(30)

$$\frac{d}{d\xi}\left[\left(\frac{1}{2}\rho u^2 + \rho \varepsilon\right)u - t_{11}u + q\right] = 0,$$

and balance laws for the constituents:

$$\frac{d}{d\xi}(\rho c_b u + J_b) = 0,$$

$$\frac{d}{d\xi}\left(\rho c_b u^2 + \frac{J_b^2}{\rho c_b} + 2u J_b + p_b\right) = \hat{m}_b,$$

(31)

$$\frac{d}{d\xi}\left\{\left[\frac{1}{2}\rho c_b\left(u + \frac{J_b}{\rho c_b}\right)^2 + \rho c_b \varepsilon_b + p_b\right]\left(u + \frac{J_b}{\rho c_b}\right)\right\} = \hat{e}_b + \hat{m}_b u,$$

where $b = 1, 2$. It may be noted that equations (30) express the conservation of fluxes of mass, momentum and energy of the mixture across the shock wave, while $(31)_1$ expresses the conservation of mass fluxes of the constituents. On the other hand, equations $(31)_{2,3}$ imply that fluxes of momenta and energies of the constituents are not conserved, but rather balanced by the mutual interactions with other constituents (source terms).

3.3. Dimensionless Shock Structure Equations

To obtain results as general as possible, we transform our problem into dimensionless form. Field variables will be scaled with respect reference values, which are taken to be ones in upstream equilibrium indicated by the subscript 0. To that end we shall introduce reference (equilibrium) mass density ρ_0, temperature T_0, and velocity $a_0 = (\gamma_0(k_B/m_0)T_0)^{1/2}$, which is the speed of sound in upstream equilibrium. Here, m_0 and γ_0 are average mass and average ratio of specific heats in upstream equilibrium. Reference length l_0 is expressed in terms of a_0 and diffusion relaxation time τ_{D12}^0 evaluated in upstream equilibrium, $l_0 = a_0 \tau_{D12}^0$. Dimensionless variables now read:

$$\tilde{\rho} = \frac{\rho}{\rho_0}, \quad \tilde{u} = \frac{u}{a_0}, \quad \tilde{T} = \frac{T}{T_0}, \quad \tilde{J}_b = \frac{J_b}{\rho_0 a_0}, \quad \tilde{\Theta}_b = \frac{\Theta_b}{T_0}, \quad \tilde{\xi} = \frac{\xi}{a_0 \tau_{D12}^0}. \tag{32}$$

Without ambiguity we may drop tilde in subsequent equations.

Mixture conservation laws for the shock structure read:

$$\frac{d}{d\xi}(\rho u) = 0,$$
$$\frac{d}{d\xi}\left(\rho u^2 - t_{11}\right) = 0, \tag{33}$$
$$\frac{d}{d\xi}\left[\left(\frac{1}{2}\rho u^2 + \rho\varepsilon\right)u - t_{11}u + q\right] = 0,$$

where we the following relations hold:

$$t_{11} = -\frac{1}{\gamma_0}\frac{m_0}{m}\rho T - \frac{1}{\rho c_1}J_1^2 - \frac{1}{\rho c_2}J_2^2 - \frac{1}{\rho c_3}J_3^2,$$
$$\rho\varepsilon = \frac{m_0}{\gamma_0}\frac{1}{(\gamma-1)m}\rho T + \frac{1}{2\rho c_1}J_1^2 + \frac{1}{2\rho c_2}J_2^2 + \frac{1}{2\rho c_3}J_3^2, \tag{34}$$
$$q = \left[\frac{1}{2\rho^2 c_1^2}J_1^2 + \frac{1}{\gamma_0}\frac{m_0}{m_1}\frac{\gamma_1}{\gamma_1-1}(T+\Theta_1)\right]J_1$$
$$+ \left[\frac{1}{2\rho^2 c_2^2}J_2^2 + \frac{1}{\gamma_0}\frac{m_0}{m_2}\frac{\gamma_2}{\gamma_2-1}(T+\Theta_2)\right]J_2$$
$$+ \left[\frac{1}{2\rho^2 c_3^2}J_3^2 + \frac{1}{\gamma_0}\frac{m_0}{m_3}\frac{\gamma_3}{\gamma_3-1}(T+\Theta_3)\right]J_3.$$

In Equation (34) we have to take into account constraints (17):

$$c_3 = 1 - c_1 - c_2, \quad J_3 = -J_1 - J_2, \quad \Theta_3 = -Y_{13}\Theta_1 - Y_{23}\Theta_2, \tag{35}$$

where Y_{13} and Y_{23} are determined by (19). Balance laws for the constituents read:

$$\frac{d}{d\xi}(\rho c_b u + J_b) = 0,$$
$$\frac{d}{d\xi}\left(\rho c_b u^2 + \frac{J_b^2}{\rho c_b} + 2u J_b + \frac{1}{\gamma_0}\frac{m_0}{m_b}\rho c_b(T+\Theta_b)\right) = \hat{m}_b, \tag{36}$$
$$\frac{d}{d\xi}\left\{\left[\frac{1}{2}\rho c_b\left(u + \frac{J_b}{\rho c_b}\right)^2 + \frac{m_0}{m_b}\frac{\gamma_b}{\gamma_0}\frac{1}{\gamma_b-1}\rho c_b(T+\Theta_b)\right]\left(u + \frac{J_b}{\rho c_b}\right)\right\} = \hat{e}_b + \hat{m}_b u$$

where $b = 1, 2$. Dimensionless source terms $\hat{m}_b$ read:

$$
\begin{aligned}
\hat{m}_1 &= -c_{20}\, c_{30} \left[\frac{\psi_{11}}{\psi_{12}^0} \left(\frac{J_1}{\rho c_1(T + \Theta_1)} - \frac{J_3}{\rho c_3(T + \Theta_3)} \right) \right. \\
&\left. + \frac{\psi_{12}}{\psi_{12}^0} \left(\frac{J_2}{\rho c_2(T + \Theta_2)} - \frac{J_3}{\rho c_3(T + \Theta_3)} \right) \right], \\
\hat{m}_2 &= -c_{20}\, c_{30} \left[\frac{\psi_{21}}{\psi_{12}^0} \left(\frac{J_1}{\rho c_1(T + \Theta_1)} - \frac{J_3}{\rho c_3(T + \Theta_3)} \right) \right. \\
&\left. + \frac{\psi_{22}}{\psi_{12}^0} \left(\frac{J_2}{\rho c_2(T + \Theta_2)} - \frac{J_3}{\rho c_3(T + \Theta_3)} \right) \right].
\end{aligned}
\tag{37}
$$

Dimensionless source terms $\hat{e}_b$ read:

$$
\begin{aligned}
\hat{e}_1 &= -\, c_{20}\, c_{30} \left[\frac{\theta_{11}}{a_0^2 \psi_{12}^0} \frac{\Theta_1 - \Theta_3}{(T + \Theta_1)(T + \Theta_3)} + \frac{\theta_{12}}{a_0^2 \psi_{12}^0} \frac{\Theta_2 - \Theta_3}{(T + \Theta_2)(T + \Theta_3)} \right], \\
\hat{e}_2 &= -\, c_{20}\, c_{30} \left[\frac{\theta_{21}}{a_0^2 \psi_{12}^0} \frac{\Theta_1 - \Theta_3}{(T + \Theta_1)(T + \Theta_3)} + \frac{\theta_{22}}{a_0^2 \psi_{12}^0} \frac{\Theta_2 - \Theta_3}{(T + \Theta_2)(T + \Theta_3)} \right].
\end{aligned}
\tag{38}
$$

In (37) and (38), c_{20} and c_{30} denote mass concentrations in upstream equilibrium. Ratios of phenomenological coefficients are given in the Appendix A.

Governing Equations (33) and (36), along with (34), (37) and (38), constitute the system of (dimensionless) shock structure equations for a multi-temperature three-component mixture of Euler fluids. If the shock structure is analyzed in a single-temperature model, then equations $(36)_3$ should be dropped from the system, and $\Theta_1 = \Theta_2 = \Theta_3 \equiv 0$ set in the remaining equations.

3.4. Parameters of the Model

The dimensionless form of the model substantially reduces the number of parameters. In previous study of binary mixtures [47] there were three parameters upon which the solution depend—Mach number and mass concentration of one constituent in upstream equilibrium, and mass ratio of the constituents. Increase of the number of constituents increases the number parameters. Thus, in this study, in which we deal with $n = 3$, constituents the following parameters will be used:

$$M_0 = \frac{u_0}{a_0}, \qquad\qquad \text{Mach number in upstream equilibrium;}$$

$$c_{10} = \left(\frac{\rho_1}{\rho} \right)_0, \; c_{20} = \left(\frac{\rho_2}{\rho} \right)_0, \qquad \text{mass concentrations in upstream equilibrium;}$$

$$\mu_1 = \frac{m_1}{m_3}, \; \mu_2 = \frac{m_2}{m_3}, \qquad \text{mass ratios of the constituents.}$$

To avoid ambiguity, we shall assume that the following inequality holds, $m_1 < m_2 < m_3$, which implies $0 < \mu_1 < \mu_2 < 1$. Additionally, the ratios of specific heats of the constituents $\gamma_\alpha, \alpha = 1, 2, 3$, have to be fixed.

3.5. Numerical Procedure

Governing equations of the shock structure consist of conservation laws (33) for the mixture, and balance laws for the constituents (36), where explicit form of the fluxes and source terms are given by (34), (37) and (38), respectively. Governing equations could be formally written in the form:

$$\frac{\mathrm{d}}{\mathrm{d}\xi} \mathbf{F}(\mathbf{U}(\xi)) = \mathbf{f}(\mathbf{U}(\xi)), \tag{39}$$

where $\mathbf{U} = (\rho, u, T, c_1, c_2, J_1, J_2, \Theta_1, \Theta_2)^T$ denotes the vector of field variables, and $\mathbf{F}(\mathbf{U})$ and $\mathbf{f}(\mathbf{U})$ have obvious meaning.

Shock structure is continuous solution of the system (39) which asymptotically connects upstream equilibrium $\mathbf{U}_-$ with downstream equilibrium $\mathbf{U}_+$, i.e., a heteroclinic orbit. Equilibrium states are not independent, but related by Rankine–Hugoniot equations. As a consequence, the system of first-order ordinary differential equations (39) is adjoined with the following boundary conditions:

$$\mathbf{U}(-\infty) = \lim_{\zeta \to -\infty} \mathbf{U}(\zeta) = \mathbf{U}_-, \quad \mathbf{U}(+\infty) = \lim_{\zeta \to +\infty} \mathbf{U}(\zeta) = \mathbf{U}_+,$$

$$\mathbf{U}'(\pm\infty) = \lim_{\zeta \to \pm\infty} \frac{\mathrm{d}\mathbf{U}(\zeta)}{\mathrm{d}\zeta} = \mathbf{0},$$

(40)

where the boundary data are as follows:

$$\mathbf{U}_- = \begin{bmatrix} \rho_- \\ u_- \\ T_- \\ c_{1-} \\ c_{2-} \\ J_{1-} \\ J_{2-} \\ \Theta_{1-} \\ \Theta_2 \end{bmatrix} = \begin{bmatrix} 1 \\ M_0 \\ 1 \\ c_{10} \\ c_{20} \\ 0 \\ 0 \\ 0 \\ 0 \end{bmatrix}, \quad \mathbf{U}_+ = \begin{bmatrix} \rho_+ \\ u_+ \\ T_+ \\ c_{1+} \\ c_{2+} \\ J_{1+} \\ J_{2+} \\ \Theta_{1+} \\ \Theta_{2+} \end{bmatrix} = \begin{bmatrix} \dfrac{M_0^2}{1-\mu_0^2+\mu_0^2 M_0^2} \\[2mm] \dfrac{1-\mu_0^2+\mu_0^2 M_0^2}{M_0} \\[2mm] \dfrac{1-\mu_0^2+\mu_0^2 M_0^2}{M_0^2}\left[(1+\mu_0^2)M_0^2 - \mu_0^2\right] \\[2mm] c_{10} \\ c_{20} \\ 0 \\ 0 \\ 0 \\ 0 \end{bmatrix},$$

(41)

where $\mu_0^2 = (\gamma_0 - 1)/(\gamma_0 + 1)$. In the case of a ST mixture, field variables Θ_1 and Θ_2 should be dropped from the vector $\mathbf{U}$, as well as from boundary data (41).

Note that the role of Rankine–Hugoniot equations in this problem is not the same as in the models without dissipation, like Euler equations for single-component gases [1]. In the Euler equations, they relate the jump of field variables at the shock wave (singular surface) to the shock speed. Here, instead, they relate the equilibrium states, upstream and downstream, which are connected by a continuous shock structure. However, they coincide with Rankine–Hugoniot equations for Euler equations because equilibrium states lie on equilibrium manifold, and the equilibrium subsystem for our model coincides with the Euler equations. They are derived by means of integration of conservation laws (33), taking into account equilibrium values of non-convective fluxes t_{11} and q from (34) and $J_\alpha = 0$.

Numerical computation of the shock structure in hyperbolic systems of balance laws is delicate because continuous solution does not exist for all values of the shock speed. Namely, it was shown that continuous solution breaks down, and there appears the profile with a sub-shock, if shock speed exceeds the greatest characteristic speed of the system in upstream equilibrium [38]. However, there may also appear a regular singularity, within the continuous profile, if the shock speed reaches the critical value in downstream equilibrium. These cases are carefully analyzed in [35], and existence of continuous shock profiles in mixtures was the subject of several recent studies [48,49].

To find numerical solution of the problem (39) (11), the infinite domain of definition of heteroclinic orbit, $-\infty < \zeta < \infty$, has to be truncated; however, the new finite domain, $\zeta \in [\zeta_0, \zeta_1]$, has to be large enough to secure that the whole profile is properly captured. Further computation may be based upon two different strategies. First one takes into account the type of stationary/equilibrium points and finds the shock structure as solution of the initial value problem. This approach is described in details in [47]. It is simple, straightforward, with low computational cost. Nevertheless, its applicability is limited since regular singularity, if it appears, cannot be overcome. Second strategy uses the finite-difference scheme to compute the shock structure as solution of the boundary value problem. This method may require several steps of computation. Adaptive step size has to be used, with smaller steps in the sub-domain in which steep gradients of field variables occur. However, position of this sub-domain is not known beforehand. Additionally, an initial guess has to be provided for numerical computation. On the other hand, boundary value approach may produce continuous solution even when regular singularity appear in the domain. Details about this approach may be found in [18,53].

Our analysis will try to take advantages of both strategies described above. In cases in which initial value problem can be solved, we shall first use this method, and then exploit this solution as initial guess for the boundary value problem. In such a way we shall get the estimate of domain width and the location of sub-domain with steep gradients in first place. After that, finite-difference scheme will be used to further improve the solution. On the other hand, when initial value strategy is not applicable, we shall use as an initial guess the continuous solution obtained for some other (not too distinct) values of parameters.

4. Analysis of the Shock Structure

Analysis of the shock structure in three-component mixture of Euler fluids has several goals. First, we want to validate our model, especially phenomenological coefficients, by computing typical profiles that are comparable with the ones obtained by other methods. Second, we want to recognize the features typical for polyatomic gases. Third, we want get a glimpse of the influence of Mach number on the shock structure. Fourth, we want set up the data which mimic the binary mixture and find out what are the differences that three-component model brings. Finally, we shall compare the multi-temperature shock structure with the single-temperature one, and try to estimate the effect obtained by the MT assumption.

Computation are performed in the way explained in Section 3. However, field variables are scaled to change in the $[0, 1]$ interval, where it is appropriate.

4.1. Shock Structure in the MT Model

Our model contains several parameters, whose values have to be given prior to numerical computation. First, phenomenological coefficients (21) depend on the type of particle interaction through the parameter $s_{\alpha\beta}$ of the cross-section, (22) and (23). In the sequel, we shall assume $s_{\alpha\beta} = 1$, which corresponds to the hard-sphere model of particle interaction. Furthermore, in all the computations we shall assume the following values of ratios of specific heats: $\gamma_1 = 7/5$, $\gamma_2 = \gamma_3 = 9/7$. In most of the computations mass ratios will be $\mu_1 = 0.01$ and $\mu_2 = 0.1$, unless stated otherwise (see Case 4). This assumption tends to cover the most interesting case from the MT point of view—the one with a mixture whose constituents have disparate masses.

Case 1. In this case it is analyzed the situation with small equilibrium concentration c_{30} of the heaviest constituent, and large equilibrium concentration c_{20}. Shock structure is computed for $c_{10} = 0.10$, $c_{20} = 0.75$ and $c_{30} = 0.15$, and $M_0 = 1.30$, and shown in Figure 1. Average mixture variables ρ, u and T have proper monotonic profiles, as well as velocities of the constituents. However, it may be observed that temperature T_3 of the heaviest constituent has an overshoot—the zone in which the variable has the value greater than the downstream equilibrium value. This result corresponds to similar observations in binary mixtures, where an overshoot appears when heavier constituents have small concentration [41,42,47]. It may also be noted that the difference between the velocities of the constituents is significant. Therefore, this case corresponds to typical profiles which can be found in [55,56]. Implicitly, it validates the structure of phenomenological coefficients (21) obtained by matching the continuum and kinetic approach.

Case 2. In the second case, two profiles were computed for the same value of Mach number, $M_0 = 1.20$. The difference appears in equilibrium concentrations: for $c_{10} = 0.60$, $c_{20} = 0.30$, and $c_{30} = 0.10$, profiles are presented in Figure 2, graphs (a) and (b); for $c_{10} = 0.60$, $c_{20} = 0.20$, and $c_{30} = 0.20$, profiles are presented in Figure 2, graphs (c) and (d). The profiles consist of two zones: upstream one with steeper gradients of field variables, and downstream one with moderate gradients. This appears to be more prominent in the profiles of mixture variables, especially on graph (a). This kind of profile, named Type-C, was firstly observed in [59] for the shock structure in ET14 model of a polyatomic gas (see also [18] for a contextualized account). The result was further supported by numerical solution of the Boltzmann equation for polyatomic gases [60], as well as two-temperature Navier–Stokes equations for polyatomic gases [61]. The main cause of such a profile is attributed to the presence of dynamic pressure and its relaxation (bulk viscosity), which is longer than the relaxation of viscous stress and heat flux. Although our model does not take into account these processes, there exists certain resemblance to this phenomenon in our profiles as well. To the best of our knowledge, such profiles do not appear in monatomic gases and their mixtures, and we are convinced that polyatomic molecular structure is the cause for this result. Note that such profiles appear in the case with an overshoot and without it alike.

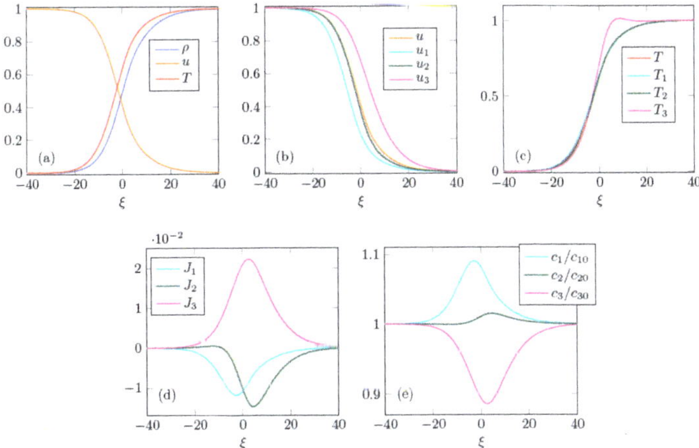

Figure 1. Shock structure for $c_{10} = 0.10$, $c_{20} = 0.75$, $c_{30} = 0.15$, and $M_0 = 1.30$: (**a**) mixture field variables; (**b**) velocities; (**c**) temperatures; (**d**) diffusion fluxes; (**e**) concentrations.

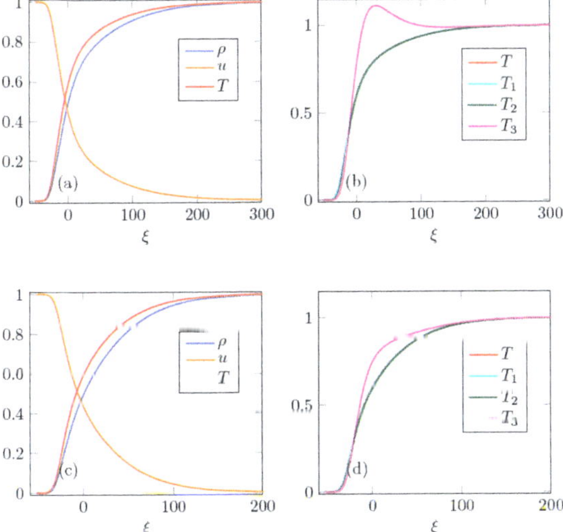

Figure 2. Shock structure for $M_0 = 1.20$, $c_{10} = 0.60$, $c_{20} = 0.30$, and $c_{30} = 0.10$. (**a**) mixture field variables; (**b**) temperatures. Shock structure for $M_0 = 1.20$, $c_{10} = 0.60$, $c_{20} = 0.20$, and $c_{30} = 0.20$: (**c**) mixture field variables; (**d**) temperatures.

Case 3. In this case we examined the profiles with the highest equilibrium concentration c_{30}. In particular, we computed the profiles for $c_{10} = 0.20$, $c_{20} = 0.20$ and $c_{30} = 0.60$, and for two different values of Mach number, $M_0 = 1.20$ and $M_0 = 1.40$, presented in Figure 3a,b, respectively. For these profiles it is remarkable that temperatures T_α of the constituents have negligibly small difference, which is almost indistinguishable on the scale of graph (and thus not presented here). Therefore, we present the two cases which are in accordance with the results for binary mixture [47]—shock thickness decreases with the increase of Mach number. This conclusion is valid for small values of M_0.

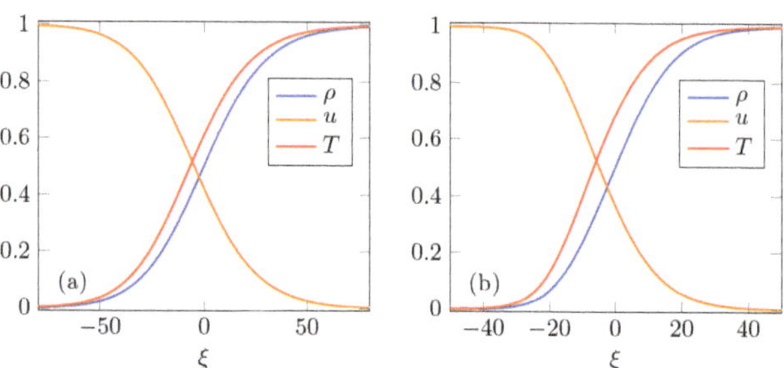

Figure 3. Shock structure for $c_{10} = 0.20$, $c_{20} = 0.20$ and $c_{30} = 0.60$: (**a**) mixture field variables for $M_0 = 1.20$; (**b**) mixture field variables for $M_0 = 1.40$.

Case 4. Here we computed the profiles for $c_{10} = 0.40$, $c_{20} = 0.50$ and $c_{30} = 0.10$. However, we analyzed two different cases: (a) $M_0 = 1.20$ and $\mu_2 = 0.10$, and (b) $M_0 = 1.40$ and $\mu_2 = 0.80$, presented in Figure 4. These results give an insight into the influence of mass ratio on the shock structure. In case (a), in the mixture with disparate masses of constituents, one may observe the temperature overshoot in T_3. However, in case (b) where constituents 2 and 3 have comparable masses, temperatures of the constituents are almost indistinguishable.

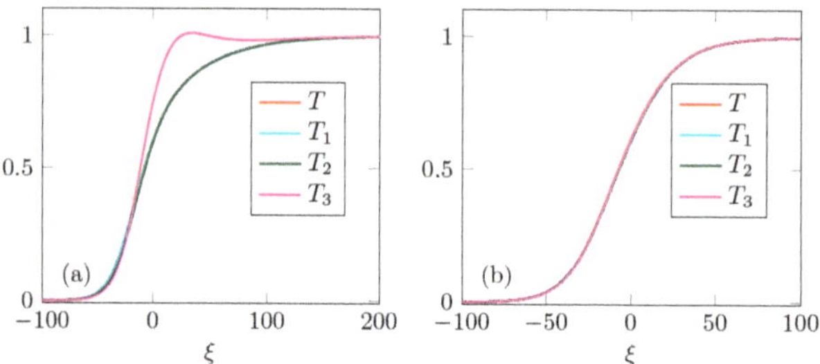

Figure 4. Shock structure for $c_{10} = 0.40$, $c_{20} = 0.50$ and $c_{30} = 0.10$: (**a**) temperatures for $M_0 = 1.20$ and $\mu_2 = 0.10$; (**b**) temperatures for $M_0 = 1.40$ and $\mu_2 = 0.80$.

4.2. Comparison with a Single-Temperature Model

There always appears a question of the relevance of the MT assumption in the mixtures. Certainly, there are non-equilibrium processes in which this assumption cannot be dropped (e.g., in plasma). On the other hand, it also proved to be relevant in mixtures whose constituents have disparate atomic/molecular masses, like noble gases. Within the framework of RET, even if the multi-temperature assumption is dropped, one ends up with a multi-velocity model which is still hyperbolic and does not produce the paradox of infinite speed of pulse propagation.

Our intention is to quantify the discrepancy in mixture field variables, primarily in the average temperature of the mixture, which appears within the shock profile computed for MT model, and the one computed for ST model. To that end, we compute the sup-norm of the difference of numerically computed solutions, i.e., $\|T - T_{\mathrm{ST}}\| = \sup_{\xi \in [\xi_0, \xi_1]} |T(\xi) - T_{\mathrm{ST}}(\xi)|$, where $T_{\mathrm{ST}}(\xi)$ denotes the temperature field along the shock profile in ST model.

For the shock structure with a temperature overshoot, presented in Case 1. of this study, sup-norm of the difference of normalized temperature profiles (the ones whose equilibrium values are set to 0.0 and 1.0) is $\|T - T_{\mathrm{ST}}\| \sim 0.01$ (see Figure 5). This implies that processes of energy exchange between the constituents, which appear due to the temperature difference, do not have a significant influence on the average value of temperature in MT model. However, this does not mean that they may be neglected. It was shown that appearance of temperature overshoot is directly related to insufficient energy exchange/transfer [43,47]. Additionally, preliminary computations, not reported here, show that there are cases with extremely large temperature overshoot, $(T_3^{\max} - T_+)/(T_+ - T_-) \approx 0.70$, in which ST is not capable of proper capturing the behavior of the mixture temperature. This will be the subject of our prospective study.

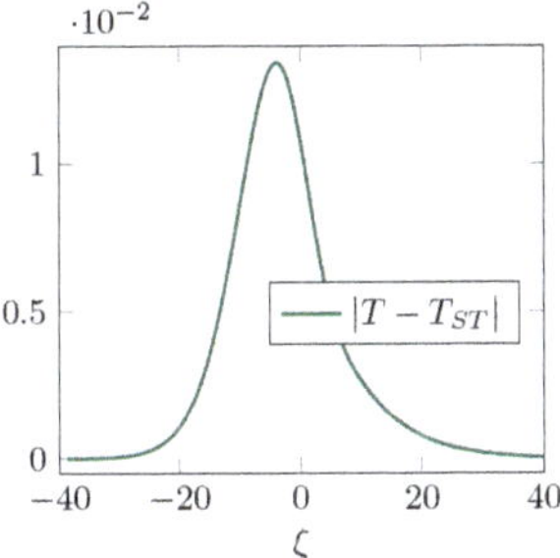

Figure 5. Difference of the mean temperatures in MT and ST model, computed for the Case 1, $c_{10} = 0.10$, $c_{20} = 0.75$, $c_{30} = 0.15$, and $M_0 = 1.30$.

5. Relaxation in the Shock Structure

One of the main features of hyperbolic systems of balance laws is that dissipation is achieved through the relaxation process, i.e., convergence of non-equilibrium variables to a certain (local) equilibrium state. Quantitative measure of the rate of convergence towards equilibrium is relaxation time. In simpler mathematical models, relaxation time may be unique for the whole system. However, if the model describes complex physical processes, there may be a need for more than one relaxation time. This is the case in our model of multi-component mixtures.

Mechanical (diffusion) and thermal relaxation times were introduced in (25), while their influence on the source terms was described in (26). It was shown recently [53] that the rate of convergence may not be the same for different processes. Therefore, our aim is to compare different relaxation times and to get an insight to the rate of convergence towards equilibrium for different processes. We shall compare the relaxation times in a relative sense, i.e., by computing the ratio of relaxation times τ_{Dbc} and τ_{Tbc} to the reference relaxation time τ_{D12}^0 along the shock profiles. After straightforward calculation, one may find the explicit dimensionless form of ratios of the diffusion relaxation times to the reference one:

$$\frac{\tau_{D11}}{\tau_{D12}^0} = \frac{\psi_{12}^0}{\psi_{11}} \frac{c_1 c_3}{c_{20} c_{30}} \rho T = \tilde{\tau}_{D11}; \qquad \frac{\tau_{D12}}{\tau_{D12}^0} = -\frac{\psi_{12}^0}{\psi_{12}} \frac{c_2 c_3}{c_{20} c_{30}} \rho T = \tilde{\tau}_{D12};$$

$$\frac{\tau_{D21}}{\tau_{D12}^0} = -\frac{\psi_{12}^0}{\psi_{21}} \frac{c_1 c_3}{c_{20} c_{30}} \rho T = \tilde{\tau}_{D21}; \qquad \frac{\tau_{D22}}{\tau_{D12}^0} = \frac{\psi_{12}^0}{\psi_{22}} \frac{c_2 c_3}{c_{20} c_{30}} \rho T = \tilde{\tau}_{D22}.$$

Ratios of the thermal relaxation times to the reference one in dimensionless form read:

$$\frac{\tau_{T11}}{\tau_{D12}^0} = \frac{a_0^2 \psi_{12}^0}{\theta_{11}} \frac{\gamma-1}{\gamma_0(\gamma_1-1)(\gamma_3-1)} \frac{m_0 m}{m_1 m_3} \frac{c_1 c_3}{c_{20} c_{30}} \rho T^2 = \tilde{\tau}_{T11};$$

$$\frac{\tau_{T12}}{\tau_{D12}^0} = -\frac{a_0^2 \psi_{12}^0}{\theta_{12}} \frac{\gamma-1}{\gamma_0(\gamma_2-1)(\gamma_3-1)} \frac{m_0 m}{m_2 m_3} \frac{c_2 c_3}{c_{20} c_{30}} \rho T^2 = \tilde{\tau}_{T12};$$

$$\frac{\tau_{T21}}{\tau_{D12}^0} = -\frac{a_0^2 \psi_{12}^0}{\theta_{21}} \frac{\gamma-1}{\gamma_0(\gamma_1-1)(\gamma_3-1)} \frac{m_0 m}{m_1 m_3} \frac{c_1 c_3}{c_{20} c_{30}} \rho T^2 = \tilde{\tau}_{T21};$$

$$\frac{\tau_{T22}}{\tau_{D12}^0} = \frac{a_0^2 \psi_{12}^0}{\theta_{22}} \frac{\gamma-1}{\gamma_0(\gamma_2-1)(\gamma_3-1)} \frac{m_0 m}{m_2 m_3} \frac{c_2 c_3}{c_{20} c_{30}} \rho T^2 = \tilde{\tau}_{T22}.$$

Ratios of the phenomenological coefficients needed to complete these expressions are given in the Appendix A.

Relaxation times, presented in Figure 6, are computed for the profiles analyzed in Case 1. They share one common property—a decrease along the shock profile, although it may not be monotonic. However, the most remarkable feature of the relaxation times in the multi-component mixture, at least in this case, is that the thermal relaxation times $\tilde{\tau}_{Tbc}$ are considerably smaller than the diffusion relaxation times $\tilde{\tau}_{Dbc}$. These results oppose the ones obtained in the case of binary mixture [40,47], even when the viscous and thermal dissipation are taken into account [53]. This certainly calls for deeper analysis of relaxation processes in the mixture. Additionally, it must be kept in mind that relaxation times may be strictly related to the rate of convergence towards equilibrium only in the simplest possible space-homogeneous cases. Therefore, relaxation times (25) in our model cannot be simply identified as the time-rates of convergence of non-equilibrium variables.

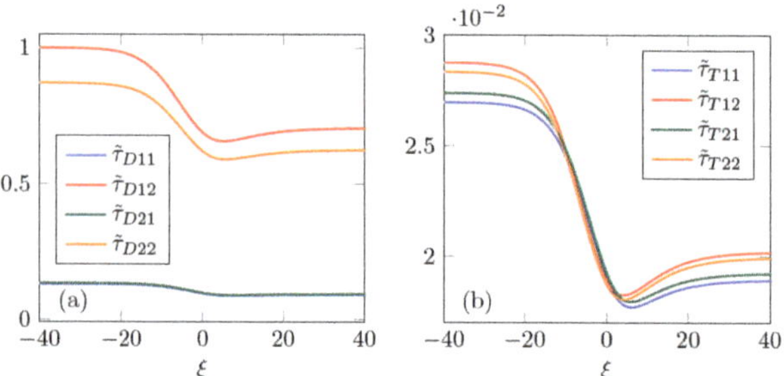

Figure 6. Relaxation times in the shock structure for $c_{10} = 0.10$, $c_{20} = 0.75$, $c_{30} = 0.15$, and $M_0 = 1.30$: (**a**) mechanical (diffusion) relaxation times; (**b**) thermal relaxation times.

6. Conclusions and Outlook

This paper was devoted to the study of shock structure in multi-component mixture of Euler fluids. Analysis was based upon the hyperbolic MT model for mixtures developed within the framework of RET. Explicit form of phenomenological coefficients, which appear in the source terms, was determined by matching with a linearized weak form of the collision operator in the system of Boltzmann equations for mixtures of polyatomic gases. The analysis was restricted to the normal shock waves, smeared out into a shock structure due to dissipation. Therefore, the shock structure was assumed in the form of a traveling wave. The original governing equations are transformed into a system of ODE's that determine the shock structure as a heteroclinic orbit which asymptotically connects stationary points. Shock profiles were found as numerical solutions to the system of shock structure equations in dimensionless form, and analyzed for the several different sets of values of the parameters.

Our attention was focused on cases which possess certain distinguishing features, and thus can be taken as representatives of a broader class of qualitatively similar shock profiles. In particular, the following representative cases were recognized:

(a) shock profiles with an overshoot in T_3 (the temperature of the heaviest constituent); this appears in the mixture with disparate masses of the constituents, $\mu_1 = 0.01$, $\mu_2 = 0.1$, and when the equilibrium concentration c_{30} of the heaviest constituent is small;

(b) shock profiles with two zones; in this case one distinguishes the front part of the profile, with steep gradients of field variables, and the rear part with moderate gradients of field variables;

(c) shock profiles with small magnitude of the diffusion temperatures, and decreasing thickness with the increase of Mach number; this case is typical for large equilibrium concentration c_{30} of the heaviest constituent;

(d) shock profiles which depend on the mass ratio; increase of the mass ratio μ_2 leads to disappearance of the temperature overshoot, and resembles the behavior of binary mixture.

We also analyzed the ST model and compared its profiles with the MT ones. It was recognized that MT assumption does not produce significant difference between the profiles of the average temperatures in MT and ST case. However, certain profiles exist with larger magnitude of temperature overshoot, and systematic analysis of MT assumption in these cases is a matter of ongoing research.

Finally, to perceive the rate at which dissipation occurs, we computed the relaxation times which appear in the source terms. It was found out that relaxation times decreased along the shock profile (but not monotonically). Analysis also shed new light on their magnitudes by giving an estimate that thermal relaxation times could be smaller than the diffusion ones.

The present study opens a new chapter in analysis of the shock structure in gaseous mixtures. It is important to note that the model took advantage of the continuum approach, which secured thermodynamic consistency and peculiar structure of the source terms, and the kinetic theory, which equipped us with explicit form of the phenomenological coefficients. In such a way we obtain reliable and tractable mathematical model at the same time. Results obtained in this way are in good agreement with the results of other studies, and therefore confirm validity of the model. Further analysis, which is a work in progress, will be concerned with a systematic study of the shock structure in the regions of parameter space in which continuous solutions exist. This will give a deeper insight into the influence of certain parameters on quantitative characteristics of the shock profiles, like the temperature overshoot or the shock thickness. Additionally, in continuation, it is our intention to extend the analysis of the entropy growth and the entropy production rate, given in [52] for the binary mixture, to three-component mixtures. Our aim is also to take into account viscous and thermal dissipation and evaluate their influence on the shock structure in multi-component mixtures, like it was done in [53] for binary mixtures. Finally, it is our intent to analyze the cases of specific mixtures. To that end we plan to use novel models of polyatomic gases, derived in the kinetic theory [21], to get better estimates of phenomenological coefficients and to relate them to measurable physical quantities like diffusivity. This will also lead to possible explicit estimate of the shock thickness in terms of the average mean free path of the molecules

Author Contributions: Conceptualization, M.P.-Č. and S.S.; methodology, D.M., M.P.-Č. and S.S.; software, D.M., formal analysis, M.P.-Č. and S.S.; investigation, D.M. and M.P.-Č.; writing—original draft preparation, M.P.-Č. and S.S.; writing—review and editing, D.M., M.P.-Č. and S.S.; visualization, D.M.; funding acquisition, M.P.-Č. and S.S. All authors have read and agreed to the published version of the manuscript.

Funding: This research was funded (D. Madjarević and M. Pavić-Čolić) by the support of the Program for Excellent Projects of Young Researchers (PROMIS) of the Science Fund of the Republic of Serbia as members of the project MaKiPol #6066089, Mathematical methods in the kinetic theory of polyatomic gas mixtures: modelling, analysis and computation, whose support enabled publication of this article in an open access manner. The research was also supported (S. Simić) by the Ministry of Education, Science and Technological Development of the Republic of Serbia (Grant No. 451-03-9/2021-14/200125).

Conflicts of Interest: The authors declare no conflict of interest.

Abbreviations

The following abbreviations are used in this manuscript:

RET   Rational extended thermodynamics
MT   Multi-temperature
ST   Single-temperature

Appendix A

Here we list the ratios of phenomenological coefficients which appear in the dimensionless form of the source terms (37) and (38). We use the abbreviated notation, $\psi_{bc} = \psi_{bc}(\mathbf{w}^0)$, $\theta_{bc} = \theta_{bc}(\mathbf{w}^0)$, and ψ_{12}^0 denotes the phenomenological coefficient evaluated at upstream equilibrium. We also take into account that coefficients are symmetric, i.e., $\psi_{21} = \psi_{12}$ and $\theta_{21} = \theta_{12}$.

Ratios of coefficients in (37) read:

$$\frac{\psi_{11}}{\psi_{12}^0} = T\left[\frac{K_{12}^\psi}{K_{12}^{\psi 0}} + \frac{\mu_{13}}{\mu_{12}}\frac{K_{13}^\psi}{K_{12}^{\psi 0}}\right], \qquad \frac{\psi_{12}}{\psi_{12}^0} = -T\frac{K_{12}^\psi}{K_{12}^{\psi 0}},$$

$$\frac{\psi_{21}}{\psi_{12}^0} = \frac{\psi_{12}}{\psi_{12}^0} = -T\frac{K_{12}^\psi}{K_{12}^{\psi 0}}, \qquad \frac{\psi_{22}}{\psi_{12}^0} = T\left[\frac{K_{12}^\psi}{K_{12}^{\psi 0}} + \frac{\mu_{23}}{\mu_{12}}\frac{K_{23}^\psi}{K_{12}^{\psi 0}}\right].$$

Ratios of coefficients in (38) read:

$$\frac{\theta_{11}}{a_0^2\psi_{12}^0} = \frac{3}{2}\frac{1}{\gamma_0}\frac{m_0}{\mu_{12}}T^2\left[\kappa_{12}^\theta\frac{K_{12}^\psi}{K_{12}^{\psi 0}} + \kappa_{13}^\theta\frac{K_{13}^\psi}{K_{12}^{\psi 0}}\right], \qquad \frac{\theta_{12}}{a_0^2\psi_{12}^0} = -\frac{3}{2}\frac{1}{\gamma_0}\frac{m_0}{\mu_{12}}T^2\kappa_{12}^\theta\frac{K_{12}^\psi}{K_{12}^{\psi 0}},$$

$$\frac{\theta_{21}}{a_0^2\psi_{12}^0} = \frac{\theta_{12}}{a_0^2\psi_{12}^0} = -\frac{3}{2}\frac{1}{\gamma_0}\frac{m_0}{\mu_{12}}T^2\kappa_{12}^\theta\frac{K_{12}^\psi}{K_{12}^{\psi 0}}, \qquad \frac{\theta_{22}}{a_0^2\psi_{12}^0} = \frac{3}{2}\frac{1}{\gamma_0}\frac{m_0}{\mu_{12}}T^2\left[\kappa_{12}^\theta\frac{K_{12}^\psi}{K_{12}^{\psi 0}} + \kappa_{23}^\theta\frac{K_{23}^\psi}{K_{12}^{\psi 0}}\right].$$

Ratios of coefficients given above are expressed in terms of K_{bc}^ψ given in (22). Taking into account (22)–(24), one may find their explicit form:

$$\frac{K_{12}^\psi}{K_{12}^{\psi 0}} = \rho^2\frac{c_1\,c_2}{c_{10}\,c_{20}}T^{-(d_1+d_2)+\frac{s_{12}}{2}},$$

$$\frac{K_{13}^\psi}{K_{12}^{\psi 0}} = \rho^2\frac{c_1\,c_3}{c_{10}\,c_{20}}\frac{\kappa_{13}^\psi}{\kappa_{12}^\psi}\mu_2 T^{-(d_1+d_3)+\frac{s_{13}}{2}},$$

$$\frac{K_{23}^\psi}{K_{12}^{\psi 0}} = \rho^2\frac{c_2\,c_3}{c_{10}\,c_{20}}\frac{\kappa_{23}^\psi}{\kappa_{12}^\psi}\mu_1 T^{-(d_2+d_3)+\frac{s_{23}}{2}}.$$

To facilitate numerical computation, we list below some useful expressions which appear in the model:

$$\frac{m_0}{m} = \left(\frac{c_1}{\mu_1} + \frac{c_2}{\mu_2} + c_3\right)\left(\frac{c_{10}}{\mu_1} + \frac{c_{20}}{\mu_2} + c_{30}\right)^{-1}, \quad \frac{m_0}{m_1} = \frac{1}{\mu_1}\left(\frac{c_{10}}{\mu_1} + \frac{c_{20}}{\mu_2} + c_{30}\right)^{-1},$$

$$\frac{m_0}{m_2} = \frac{1}{\mu_2}\left(\frac{c_{10}}{\mu_1} + \frac{c_{20}}{\mu_2} + c_{30}\right)^{-1}, \quad \frac{m_0}{m_3} = \left(\frac{c_{10}}{\mu_1} + \frac{c_{20}}{\mu_2} + c_{30}\right)^{-1},$$

$$\frac{m_0}{\mu_{12}} = \frac{\mu_1 + \mu_2}{\mu_1\mu_2}\left(\frac{c_{10}}{\mu_1} + \frac{c_{20}}{\mu_2} + c_{30}\right)^{-1},$$

$$\frac{\mu_{13}}{\mu_{12}} = \frac{\mu_1}{\mu_1 + 1}\frac{\mu_1 + \mu_2}{\mu_1\mu_2}, \quad \frac{\mu_{23}}{\mu_{12}} = \frac{\mu_2}{\mu_2 + 1}\frac{\mu_1 + \mu_2}{\mu_1\mu_2}.$$

References

1. Courant, R.; Friedrichs, K.O. *Supersonic Flow and Shock Waves*; Springer: New York, NY, USA, 1976; Volume 21.
2. Zel'dovich, Y.B; Raizer, Y.P. *Physics of Shock Waves and High-Temperature Hydrodynamic Phenomena*; Dover Publications Inc.: New York, NY, USA, 2002.
3. Lax, P. Shock waves and entropy. In *Contributions to Nonlinear Functional Analysis*; Academic Press: Cambridge, MA, USA, 1971; pp. 603–634.
4. Dafermos, C. *Hyperbolic Conservation Laws in Continuum Physics*; Springer: Berlin/Heidelberg, Germany, 2016.
5. Liu, I.-S. *Continuum Mechanics*; Springer: Berlin/Heidelberg, Germany, 2002.
6. Müller I. *Thermodynamics*; Pitman: London, UK, 1985.
7. Chen, G.-Q.; Levermore, C.D.; Liu, T.-P. Hyperbolic conservation laws with stiff relaxation terms and entropy. *Commun. Pure Appl. Math.* **1994**, *47*, 787–830. [CrossRef]
8. Ikenberry, E.; Turesdell C. On the Pressures and the Flux of Energy in a Gas according to Maxwell's Kinetic Theory, I. *J. Rational Mech. Anal.* **1956**, *5*, 1–54. [CrossRef]
9. Yang, Z.; Yong W.-A. Validity of the Chapman–Enskog expansion for a class of hyperbolic relaxation systems. *J. Differ. Equ.* **2015**, *258*, 2745–2766. [CrossRef]
10. Ruggeri, T. Can constitutive relations be represented by non-local equations? *Q. Appl. Math.* **2012**, *70*, 597–611. [CrossRef]
11. Kogan, M.N. *Rarefied Gas Dynamics*; Plenum Press: New York, NY, USA, 1969.
12. Grad, H. On the kinetic theory of rarefied gases. *Commun. Pure Appl. Math.* **1949**, *2*, 331–407. [CrossRef]
13. Müller, I.; Ruggeri, T. *Rational Extended Thermodynamics*; Springer: New York, NY, USA, 1998.
14. Boudin, L.; Grec, B.; Salvarani, F. A mathematical and numerical analysis of the Maxwell-Stefan diffusion equations. *Discrete Contin. Dyn. Syst. Ser. B* **2012**, *17*, 1427–1440. [CrossRef]
15. Boudin, L.; Grec, B.; Pavan, V. The Maxwell-Stefan diffusion limit for a kinetic model of mixtures with general cross sections. *Nonlinear Anal.* **2017**, *159*, 40–61. [CrossRef]
16. Salvarani, F.; Soares, A.J. On the relaxation of the Maxwell-Stefan system to linear diffusion. *Appl. Math. Lett.* **2018** *85*, 15–21. [CrossRef]
17. Ruggeri, T.; Simić, S. On the hyperbolic system of a mixture of Eulerian fluids: A comparison between single-and multi-temperature models. *Math. Methods Appl. Sci.* **2007**, *30*, 827–849. [CrossRef]
18. Ruggeri, T.; Sugiyama, M. *Rational Extended Thermodynamics Beyond the Monatomic Gas*; Springer: New York, NY, USA, 2015.
19. Bourgat, J.-F.; Desvillettes, L.; Le Tallec P.; Perthame B. Microreversible collisions for polyatomic gases. *Eur. J. Mech. B/Fluids* **1994**, *13*, 237–254.
20. Desvillettes, L.; Monaco, R.; Salvarani F. A kinetic model allowing to obtain the energy law of polytropic gases in the presence of chemical reactions. *Eur. J. Mech. B/Fluids* **2005**, *24*, 219. [CrossRef]
21. Djordjić, V.; Pavić-Čolić, M.; Spasojević, N. Polytropic gas modelling at kinetic and macroscopic levels. *Kinet. Relat. Models* **2021**, doi:10.3934/krm.2021013. [CrossRef]
22. Dellacherie, S. On the Wang Chang-Uhlenbeck equations. *Discrete Contin. Dyn. Syst. Ser. B* **2003**, *3*, 229–253.
23. Giovangigli, V. *Multicomponent Flow Modeling*; Birkhäuser: Boston, MA, USA, 1999.
24. Nagnibeda E.; Kustova E. *Non-Equilibrium Reacting Gas Flows: Kinetic Theory of Transport and Relaxation Processes*; Springer: Berlin/Heidelberg, Germany, 2009.
25. Kremer G.M. *An Introduction to the Boltzmann Equation and Transport Processes in Gases*; Springer: Berlin/Heidelberg, Germany, 2010.
26. Pavić, M.; Ruggeri, T.; Simić, S. Maximum entropy principle for rarefied polyatomic gases. *Physica A* **2013**, *392*, 1302–1317. [CrossRef]
27. Pavić-Čolić, M.; Simić, S. Moment equations for polyatomic gases. *Acta Appl. Math.* **2014**, *132*, 469–482. [CrossRef]

28. Baranger, C.; Bisi, M.; Brull, S.; Desvillettes, L. On the Chapman-Enskog asymptotics for a mixture of monoatomic and polyatomic rarefied gases. *Kinet. Relat. Models* **2018**, *11*, 821–858. [CrossRef]
29. Anwasia, B.; Bisi, M.; Salvarani, F.; Soares, A.J. On the Maxwell-Stefan diffusion limit for a reactive mixture of polyatomic gases in non-isothermal setting (English summary). *Kinet. Relat. Models* **2020**, *13*, 63–95. [CrossRef]
30. Gamba I.M.; Pavić-Čolić, M. On the Cauchy problem for Boltzmann equation modelling a polyatomic gas. *arXiv* **2020**, arXiv:2005.01017.
31. Bisi, M.; Martalò G.; Spiga G. Multi-temperature Euler hydrodynamics for a reacting gas from a kinetic approach to rarefied mixtures with resonant collisions. *EPL (Europhys. Lett.)* **2011**, *95*, 55002. [CrossRef]
32. Bisi, M.; Groppi, M.; Martalò, G. Macroscopic equations for inert gas mixtures in different hydrodynamic regimes. *J. Phys. A Math. Theor.* **2021**, *54*, 085201. [CrossRef]
33. Pavić-Čolić, M. Multi-velocity and multi-temperature model of the mixture of polyatomic gases issuing from kinetic theory. *Phys. Lett. A* **2019**, *383*, 2829–2835. [CrossRef]
34. Gilbarg, D.; Paolucci, D. The structure of shock waves in the continuum theory of fluids. *J. Ration. Mech. Anal.* **1953**, *2*, 617–642. [CrossRef]
35. Weiss, W. Continuous shock structure in extended thermodynamics. *Phys. Rev. E* **1995**, *52*, R5760. [CrossRef] [PubMed]
36. Achleitner, F.; Szmolyan, P. Saddle–node bifurcation of viscous profiles. *Phys. D Nonlinear Phenom.* **2012**, *241*, 1703–1717. [CrossRef] [PubMed]
37. Simić, S. Shock structure in continuum models of gas dynamics: Stability and bifurcation analysis. *Nonlinearity* **2009**, *22*, 1337. [CrossRef]
38. Boillat, G.; Ruggeri, T. On the shock structure problem for hyperbolic system of balance laws and convex entropy. *Contin. Mech. Thermodyn.* **1998**, *10*, 285–292. [CrossRef]
39. Sherman, F.S. Shock-wave structure in binary mixtures of chemically inert perfect gases. *J. Fluid Mech.* **1960**, *8*, 465–480. [CrossRef]
40. Bird, G. The structure of normal shock waves in a binary gas mixture. *J. Fluid Mech.* **1968**, *31*, 657. [CrossRef]
41. Kosuge, S.; Aoki, K.; Takata, S. Shock-wave structure for a binary gas mixture: Finite-difference analysis of the Boltzmann equation for hard sphere molecules. *Eur. J. Mech. B Fluids* **2001**, *20*, 87–126. [CrossRef]
42. Raines, A.A. Study of a shock wave structure in gas mixtures on the basis of the Boltzmann equation. *Eur. J. Mech. B Fluids* **2002**, *21*, 599–610. [CrossRef]
43. Abe, K.; Oguchi, H. Shock wave structures in binary gas mixtures with regard to temperature overshoot. *Phys. Fluids* **1974**, *17*, 1333–1334. [CrossRef]
44. Bisi M.; Martalò G.; Spiga G. Shock wave structure of multi-temperature Euler equations from kinetic theory for a binary mixture. *Acta Appl. Math.* **2014**, *132*, 95–105. [CrossRef]
45. Bisi, M.; Groppi M.; Martalò, G. On the shock thickness for a binary gas mixture. *Ric. Mat.* **2020**, in press. [CrossRef]
46. Madjarević, D.; Simić, S. Shock structure in helium-argon mixture—A comparison of hyperbolic multi-temperature model with experiment. *EPL (Europhys. Lett.)* **2013**, *102*, 44002. [CrossRef]
47. Madjarević, D.; Ruggeri T.; Simić S. Shock structure and temperature overshoot in macroscopic multi-temperature model of mixtures. *Phys. Fluids* **2014**, *26*, 106102. [CrossRef]
48. Conforto, F.; Mentrelli, A.; Ruggeri, T. Shock structure and multiple sub-shocks in binary mixtures of Eulerian fluids. *Ric. Mat.* **2017**, *66*, 221–231. [CrossRef]
49. Artale, V.; Conforto, F.; Martalò, G.; Ricciardello, A. Shock structure and multiple sub-shocks in Grad 10-moment binary mixtures of monoatomic gases. *Ric. Mat.* **2019**, *68*, 485–502. [CrossRef]
50. Bisi, M.; Groppi, M.; Macaluso, A.; Martalò, G. Shock wave structure of multi-temperature Grad 10-moment equations for a binary gas mixture. *EPL (Europhys. Lett.)* **2021**, *133*, 54001. [CrossRef]
51. Sharipov, F.; Dias F.C. Structure of planar shock waves in gaseous mixtures based on ab initio direct simulation. *Eur. J. Mech. B/Fluids* **2018**, *72*, 251–263. [CrossRef]
52. Madjarević, D.; Simić, S. Entropy growth and entropy production rate in binary mixture shock waves. *Phys. Rev. E* **2019**, *100*, s023119. [CrossRef] [PubMed]
53. Simić, S.; Madjarević, D. Shock structure and entropy growth in a gaseous binary mixture with viscous and thermal dissipation. *Wave Motion* **2021**, *100*, 102661. [CrossRef]
54. Ruyev, G.A.; Fedorov A.V.; Fomin V.M. Special features of the shock-wave structure in mixtures of gases with disparate molecular masses. *J. Appl. Mech. Tech. Phys.* **2002**, *43*, 529–537. [CrossRef]
55. Josyula, E.; Vedula, P.; Bailey W.F.; Suchyta, C.J., III. Kinetic solution of the structure of a shock wave in a nonreactive gas mixture. *Phys. Fluids* **2011**, *23*, 017101. [CrossRef]
56. Raines, A.A. Numerical solution of the Boltzmann equation for the shock wave in a gas mixture. *arXiv* **2014**, arXiv:1403.3230.
57. Truesdell, C. *Rational Thermodynamics*; Springer: New York, NY, USA, 1984.
58. Boillat, G.; Ruggeri T. Hyperbolic principal subsystems: Entropy convexity and subcharacteristic conditions. *Arch. Ration. Mech. Anal.* **1997**, *137*, 305–320. [CrossRef]
59. Taniguchi, S.; Arima, T.; Ruggeri, T.; Sugiyama, M. Effect of the dynamic pressure on the shock wave structure in a rarefied polyatomic gas. *Phys. Fluids* **2014**, *26*, 016103. [CrossRef]

60. Kosuge, S.; Aoki, K. Shock-wave structure for a polyatomic gas with large bulk viscosity. *Phys. Rev. Fluids* **2018**, *3*, 023401. [CrossRef]
61. Aoki, K.; Bisi M.; Groppi, M; Kosuge S. Two-temperature Navier-Stokes equations for a polyatomic gas derived from kinetic theory. *Phys. Rev. E* **2020**, *102*, 023104. [CrossRef] [PubMed]

 *symmetry*

Article

Cyclic Control Optimization Algorithm for Stirling Engines

Raphael Paul * and Karl Heinz Hoffmann

Institute of Physics, Chemnitz University of Technology, 09107 Chemnitz, Germany;
hoffmann@physik.tu-chemnitz.de
* Correspondence: raphael.paul@physik.tu-chemnitz.de

Abstract: The ideal Stirling cycle describes a specific way to operate an equilibrium Stirling engine. This cycle consists of two isothermal and two isochoric strokes. For non-equilibrium Stirling engines, which may feature various irreversibilities and whose dynamics is characterized by a set of coupled ordinary differential equations, a control strategy that is based on the ideal cycle will not necessarily yield the best performance—for example, it will not generally lead to maximum power. In this paper, we present a method to optimize the engine's piston paths for different objectives; in particular, power and efficiency. Here, the focus is on an indirect iterative gradient algorithm that we use to solve the cyclic optimal control problem. The cyclic optimal control problem leads to a Hamiltonian system that features a symmetry between its state and costate subproblems. The symmetry manifests itself in the existence of mutually related attractive and repulsive limit cycles. Our algorithm exploits these limit cycles to solve the state and costate problems with periodic boundary conditions. A description of the algorithm is provided and it is explained how the control can be embedded in the system dynamics. Moreover, the optimization results obtained for an exemplary Stirling engine model are discussed. For this Stirling engine model, a comparison of the optimized piston paths against harmonic piston paths shows significant gains in both power and efficiency. At the maximum power point, the relative power gain due to the power-optimal control is ca. 28%, whereas the relative efficiency gain due to the efficiency-optimal control at the maximum efficiency point is ca. 10%.

Keywords: cyclic optimal control; finite time thermodynamics; endoreversible thermodynamics; Stirling; optimization

Citation: Paul, R.; Hoffmann, K.H. Cyclic Control Optimization Algorithm for Stirling Engines. *Symmetry* **2021**, *13*, 873. https://doi.org/10.3390/sym13050873

Academic Editors: Robert Kovacs, Patrizia Rogolino and Francesco Oliveri

Received: 15 April 2021
Accepted: 9 May 2021
Published: 13 May 2021

1. Introduction

Stirling engines are closed-cycle regenerative heat engines that harness a temperature difference between two external heat baths. At the cost of taking entropy from the hot heat bath and disposing it to the cold heat bath, they generate mechanical work. This means that Stirling engines are very flexible regarding the possible technical representations of those two heat baths. Therefore, they can be applied in various scenarios. Some current example applications are power generation from burnable industrial waste gases [1], domestic combined heat and power generation [2], electro-thermal energy storage systems for renewable energies as an alternative to lithium-ion batteries [3], and low maintenance power generation in remote regions with harsh climatic conditions [4], among others. If properly designed, Stirling engines can achieve high efficiency, durability, low-maintenance, and low-noise operation. Hence, they are considered to be one good candidate technology for enhancing the sustainability of power systems.

One certain ideal representation of the thermodynamic cycle performed in Stirling engines is commonly referred to as the ideal Stirling cycle. It consists of four distinct strokes (processes) that an enclosed working gas cyclically performs:

1. Isothermal compression at T_{ExL}
2. Isochoric regenerative heating to T_{ExH}
3. Isothermal expansion at T_{ExH}
4. Isochoric regenerative cooling to T_{ExL}

where T_{ExH} and T_{ExL} are the temperatures of the hot and cold heat bath, respectively. If those four ideal processes are repeated in the prescribed order, the cycle represents an ideal heat engine, operating with Carnot efficiency $\eta_C = 1 - T_{\mathrm{ExL}}/T_{\mathrm{ExH}}$. In the two regenerative processes 4 and 2, heat needs to be stored and provided at a range of temperatures. The combined process is called regeneration, and is accomplished with help of a multi-temperature heat storage, referred to as a regenerator.

In real Stirling engines, the working gas is typically contained in two working spaces. One of them, the hot working space, is in thermal contact with the hot heat bath. The other one, the cold working space, is in thermal contact with the cold heat bath. The two working spaces' volumes can be changed by moving pistons. As a result of those pistons' movements, the working spaces exchange working gas through the regenerator, which is often implemented as a porous metal structure.

In contrast to the ideal Stirling cycle, in real Stirling engines, the four above-described strokes can typically not be clearly distinguished. This is due to several reasons. First of all, the engine's piston movements (piston paths) usually do not exactly reproduce the two isochoric strokes. Moreover, the regenerator, as well as the hot and cold working spaces, have non-zero dead volumes. The most important reason, however, is connected to the fact that real engines are supposed to produce a finite amount of work in finite time, which is in contrast to the assumptions of equilibrium thermodynamics, and thus to the ideal Stirling cycle. Consequently, in real Stirling engines, various inevitable loss phenomena occur, for example, finite heat transfer between the working gas and the external heat baths or the regenerator matrix, pressure drop across the regenerator, and heat leaks through the engine's internal components.

Correspondingly, real Stirling engines are non-equilibrium devices. Hence, piston paths aiming to emulate the ideal Stirling cycle's four strokes would not necessarily yield best performance. The question of what piston paths yield best performance, for example, optimal net power, constitutes a cyclic optimal control problem.

In this paper we present an algorithm that can be used to solve such cyclic optimal control problems. It is an indirect iterative gradient algorithm and based on Optimal Control Theory. The algorithm starts with prescribing initial control functions determining the piston paths. These control functions are then gradually modified over the course of the iterations so as to gradually improve an objective and approach the optimal control. To determine the gradual shifts in every iteration, not only does the system of ordinary differential equations, describing the thermodynamics of the Stirling engine, need to be solved, but also a conjugate system of differential equations. For the efficient practical feasibility of this task, low numerical effort of the utilized thermodynamic model is crucial. In detailed models of Stirling engines, significant numerical effort is usually connected to the description of the regenerator. Therefore, here we will apply a Stirling engine model with a reduced-order endoreversible regenerator model that provides a proper tradeoff between accuracy and numerical effort for optimal control problems.

As indicated above, the core discrepancy between the ideal Stirling cycle and a real Stirling engine is that, in the former, all processes are quasi-static, and hence reversible, whereas the latter is required to produce a finite amount of work in finite time. Therefore, real Stirling engines are non-equilibrium devices that could by no means achieve the performance measures of the ideal cycle, such as Carnot efficiency. A non-equilibrium thermodynamics field that studies how constraints on time and rate influence the performance of such devices is Finite-Time Thermodynamics (FTT) [5,6]. FTT typically approaches this and related questions with variational principles and global—rather than local—descriptions of the irreversible systems considered [5].

Endoreversible Thermodynamics [7–11] can be regarded as a subfield of FTT that intends to provide a toolbox of model building blocks from which an FTT model can be constructed in a comprehensible way. The emphasis is placed on including the main loss phenomena, while keeping the model structure clear and minimizing mathematical complexity and numerical effort. The basic approach is to decompose the considered thermodynamic system into a network of reversible subsystems that are connected with each other through (ir)reversible interactions. That is, in its endoreversible representation, the original system's irreversibilities are captured by the interactions, whereas for the reversible subsystems the known relations from equilibrium thermodynamics hold.

In the past, various endoreversible systems have been subject to investigations involving control optimizations. In particular, for heat engines, the piston trajectories can be optimized for objectives, such as power or efficiency. For example, such studies have been carried out for engines with Diesel [11–14] and Otto [15–17] cycles, light-driven engines [18–20], and Stirling engines [21–23].

The optimization approach used in the current study to optimize the Stirling engine differs from [21–23] in that a cyclic version of Optimal Control Theory is applied. Beyond Endoreversible Thermodynamics, this has been done by Craun and Bamieh [24] for an ideal beta-Stirling engine model with an actively controlled displacer. In [24], the heat transfer between the heat baths and the working gas was infinitely fast and regeneration was perfect. The pressure drop across the regenerator was thermodynamically neglected, but considered in terms of mechanical losses. As opposed to this, in the current study, an alpha-Stirling engine model with finite heat transfer between the heat baths and the working gas, an external heat leak, gas leakages and friction of the piston rings, and an irreversible regenerator is considered. Compared to the endoreversible regenerator models used in [22,23], the *EEn*-regenerator model [25] used in this study features a significantly higher degree of detail.

2. Stirling Engine Model

In this paper, our focus is on a cyclic control optimization algorithm that can be applied to various Stirling engine models, or models of other cyclically operating systems. Nevertheless, we here briefly introduce a Stirling engine model [25] that we will use to demonstrate the algorithm's usage. In particular, we will present optimization results for this model in Section 5.

We consider an alpha-type Stirling engine as schematized in Figure 1, where the heat transfer between the external heat baths and the working gas is realized through the cylinders.

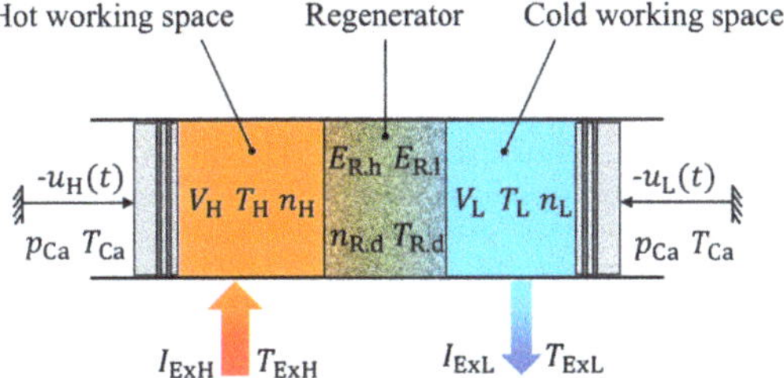

Figure 1. Schematics of the alpha-type Stirling engine considered. The corresponding endoreversible model [25] has ten degrees of freedom (state variables), being the working space's volumes V_i, temperatures T_i, and particle numbers n_i, the energies $E_{R.h}$ and $E_{R.l}$ of the hot and cold halves of the regenerator matrix as well as the particle number $n_{R.d}$ and temperature $T_{R.d}$ of the gas inside the regenerator dead space.

For the sake of simplicity, we assume a constant finite heat transfer coefficient. Further irreversibilities result from an external heat leak between the heat baths, gas leakage of the piston rings, friction of the pistons, pressure drop across the regenerator, and thermal mixing at the interfaces of the regenerator and the working spaces.

The regenerator is described with the reduced order endoreversible *EEn*-regenerator model, developed in [25]. Essentially, this is based on the assumptions that the spatial temperature distribution inside the regenerator is linear and that the temperature difference between the gas and matrix is small. Aside from external irreversibilities that automatically occur due to thermal mixing in the regenerator's interactions with the working spaces, internal irreversibilities can be included via entropy source terms in the *EEn*-regenerator model. In the current study, the irreversibility due to the pressure drop across the regenerator is accounted for by such an entropy source term.

The Stirling engine performance measures that are to be optimized in this study are power and efficiency. The power, or net power output, is defined as

$$P = \frac{W}{\tau},$$

(1)

with the fixed cycle time τ and the net cycle work

$$W = \int_0^\tau \left(p_H \dot{V}_H + p_L \dot{V}_L \right) - \gamma \left(\dot{V}_H^2 + \dot{V}_L^2 \right) \, dt,$$

(2)

where V_i and p_i are the volume and pressure of the working spaces $i \in \{H, L\}$ and γ is the friction coefficient of the pistons. Then, the efficiency is

$$\eta = \frac{\int_0^\tau \left(p_H \dot{V}_H + p_L \dot{V}_L \right) - \gamma \left(\dot{V}_H^2 + \dot{V}_L^2 \right) \, dt}{\int_0^\tau I_{H,ExH} + K_{Ex} \left(T_{ExH} - T_{ExL} \right) \, dt},$$

(3)

where $I_{H,ExH}$ is the instantaneous heat flux to the hot working space and $K_{Ex} \left(T_{ExH} - T_{ExL} \right)$ corresponds to an external heat leak. We here use the same transfer laws and parameter values as in [25], for example, the heat bath temperatures are defined as $T_{ExH} = 500$ K and $T_{ExL} = 300$ K. Solely for the heat conductance of the external heat leak, we chose a different parameter value: $K_{Ex} = 5$ W/K.

This Stirling engine model has ten state variables, which are the working spaces' volumes V_i, temperatures T_i, and particle numbers n_i with $i \in \{H, L\}$, the energies $E_{R.h}$ and $E_{R.l}$ of the hot and cold halves of the regenerator matrix as well as the particle number $n_{R.d}$ and temperature $T_{R.d}$ of the gas inside the regenerator dead space.

The state variables of this Stirling engine model can be arranged in a state vector x and the state dynamics can be expressed in terms of $\dot{x} = f(x, u)$. Here, $f(x, u)$ is a vector-valued function that depends on the state vector x and a control vector u. In our Stirling engine model we define the control vector u in terms of an explicitly time-dependent, τ-periodic control function $u(t)$, which has the entries $u_H(t)$ and $u_L(t)$. Each of those two sub-functions determines the dynamics of one working volume, as indicated in Figure 1. An essential requirement that we raise is that the state dynamics $\dot{x} = f(x, u(t))$ features a limit cycle with regard to all state variables, which will be necessary for the applicability of the optimization algorithm. Therefore, care must be taken regarding the embedding of the control in the system dynamics, which will be explained in Section 4.

3. Cyclic Optimal Control Problem

Optimal Control Theory provides a framework for the dynamic (or indirect) optimization of dynamical systems, such as the above-described Stirling engine model. The type of optimal control problem considered in this work can be formulated as: find an unconstrained, smooth, τ-periodic control function $u(t)$ that maximizes the objective functional

$$J = \int_0^\tau \zeta(x, u(t)) \, dt$$

(4)

subject to the constraint

$$\dot{x} = f(x, u(t))$$

(5)

for a fixed period τ. Here, x is the state vector of the system. The dynamics of the system is determined by the ordinary differential Equation (5) and is influenced by the vector-valued, explicitly time-dependent, τ-periodic control function $u(t)$.

For the formulation of necessary conditions of optimality, a Hamilton function H is defined as [26,27]

$$H(x, u, \lambda) := \zeta(x, u) + \lambda^T f(x, u) \tag{6}$$

with an adjoint vector λ, which varies over time and will henceforth also be called a costate vector. As we will see later, there exists an interesting symmetry between the dynamics of the state vector x and the costate vector λ. The first order necessary conditions for the control u with the optimal value are [26,27]

$$\dot{x} = \nabla_\lambda H(x, u, \lambda), \tag{7}$$

$$\dot{\lambda} = -\nabla_x H(x, u, \lambda), \tag{8}$$

$$0 = \nabla_u H(x, u, \lambda), \tag{9}$$

combined with the respective transversality conditions that specify the boundary conditions for the state and costate dynamics from Equations (7) and (8), respectively. In the case of the optimal cyclic regime, as valid for the stationary operational state of a cyclically operating engine, these boundary conditions are [27]

$$x|_\tau = x|_0, \tag{10}$$

$$\lambda|_\tau = \lambda|_0. \tag{11}$$

The notation $x|_t$ refers to the value of x at time t, while it does not indicate explicit time dependence. These boundary conditions do not render the problem overdetermined, since the initial and final values are not quantified, but only required to be equal.

In this way, the global dynamic optimization problem from Equation (4) with the constraint Equation (5) is turned into the continuous set of local (static) optimization problems from Equation (9) with the constraints Equations (7) and (8). This means that $H(x, u, \lambda)$ must be locally maximized for all $t \in [0, \tau)$ with respect to u. For the optimal solution x^*, u^*, λ^* of the considered kind of cyclic optimization problem, without explicit time dependence of $\zeta(x, u)$, the temporal value of the Hamilton function H^* will then be constant [26,27].

3.1. Maximum Power

In the case of the optimization being performed for the objective functional, which is the cycle-averaged net power output P, as defined in Equation (1), the definition of the path target function $\zeta(x, u)$ for Equations (4) and (6) is straightforward:

$$\zeta_W(x, u) := (p_H \dot{V}_H + p_L \dot{V}_L) - \gamma \left(\dot{V}_H^2 + \dot{V}_L^2\right), \tag{12}$$

which is simply the instantaneous power output. Since the optimization is performed for fixed cycle time τ, maximum power corresponds to maximum work. Hence, a prefactor of $1/\tau$ can be disregarded here.

3.2. Maximum Efficiency

In the case of the optimization being performed for the objective of efficiency η, as defined in Equation (3), the definition of $\zeta(x, u)$ is slightly more involved. This is because the definition of the efficiency corresponds to a ratio of two functionals (integrals) which need to be separately evaluated before calculating their ratio. This can be seen in Equation (3) and is in contrast to the structure of Equation (4). In order to resolve this discrepancy, we investigate how variations in the cycle work W and the heat $Q_H := \int_0^\tau I_{H,ExH} + K_{Ex}(T_{ExH} - T_{ExL})\, dt$ translate to the first variation of the efficiency:

$$\delta\eta = \delta\left(\frac{W}{Q_H}\right) = \frac{\partial\eta}{\partial W}\delta W + \frac{\partial\eta}{\partial Q_H}\delta Q_H = \frac{1}{Q_H}\delta W - \frac{W}{Q_H^2}\delta Q_H. \tag{13}$$

This means that the problem of optimizing the efficiency can be morphed into an equivalent optimization problem of the weighted sum $\lambda_W W + \lambda_{QH} Q_H$ with the weights being defined according to the partial derivatives:

$$\lambda_W = \partial\eta/\partial W = 1/Q_H \quad \text{and} \quad \lambda_{QH} = \partial\eta/\partial Q_H = -W/Q_H^2. \tag{14}$$

Within these weights, W and Q_H are the results obtained for a specific prescribed τ-periodic control function $u^\dagger(t)$ for which, at a given time t, the values of the control and state vectors are $u \approx u^\dagger(t)$ and $x \approx x^\dagger(t)$. Here, $x^\dagger(t)$ is the periodic solution of the state dynamics for $u^\dagger(t)$. Based on this, we can now define the path target function for maximizing the efficiency:

$$\zeta_\eta(x, u) := \frac{W}{Q_H\,\tau} + \frac{1}{Q_H}\zeta_W(x, u) - \frac{W}{Q_H^2}I_{H,ExH}(x, u). \tag{15}$$

Here, W and Q_H are again the results obtained for a τ-periodic control function $u^\dagger(t)$ with the above-described closeness requirements for the instantaneous values of x and u. In a strict sense, the path target function $\zeta_\eta(x, u)$ is therefore also a functional of $u^\dagger(t)$ and $x^\dagger(t)$, that is: $\zeta_\eta(x, u) = \zeta_\eta[x^\dagger(\cdot), u^\dagger(\cdot)](x, u)$ because $W = W[x^\dagger(\cdot), u^\dagger(\cdot)]$ and $Q_H = Q_H[x^\dagger(\cdot), u^\dagger(\cdot)]$. If the closeness requirement of x and u, regarding $u^\dagger(t)$ is valid for all t, the second and third terms in Equation (15) will (approximately) cancel upon integration from 0 to τ. Therefore, the first term was added here to ensure that $\int_0^\tau \zeta_\eta(x, u)\, dt$ does actually represent η. This first term does, in fact, not influence the optimization problem from Equation (7) to Equation (11), since it does not depend on the instantaneous values of x and u.

3.3. Penalty Function

The cyclic optimal control problem introduced so far does not contain inequality constraints for the state x or the control u. However, we can introduce inequality constraints for x and u by means of a penalty function. In particular, we will use such a penalty function to introduce lower and upper bounds for the Stirling engine's working volumes:

$$\zeta_{penalty}(x, u) := -v_0\left(e^{v_1\frac{V_{H,min}-V_H}{V_{H,sw}}} + e^{v_1\frac{V_H-V_{H,max}}{V_{H,sw}}} + e^{v_1\frac{V_{L,min}-V_L}{V_{L,sw}}} + e^{v_1\frac{V_L-V_{L,max}}{V_{L,sw}}}\right) \tag{16}$$

where the working volumes V_i are contained in x, $v_0 = 1\,\text{W}$ and $v_1 = 500$ are penalty factors, $V_{i,sw} = V_{i,max} - V_{i,min}$ refers to the admissible swept volumes and $V_{i,max}$ and $V_{i,min}$ are the maximum and minimum admissible working volumes with $i \in \{H, L\}$.

Accordingly, in the case of optimizing the Stirling engine's piston paths for maximum power, the overall path target function is defined as

$$\zeta(x, u) := \zeta_W(x, u) + \zeta_{penalty}(x, u), \tag{17}$$

whereas in case of the optimization for maximum efficiency, it is defined as

$$\zeta(x, u) := \zeta_\eta(x, u) + \zeta_{penalty}(x, u). \tag{18}$$

4. Optimization Algorithm

The indirect iterative gradient method described in the following was inspired by a contribution of Craun and Bamieh [24,28], who solved an optimal control problem of a modified beta-Stirling engine using an ideal thermodynamic model, namely the Schmidt model. Similar methods have earlier been used, for example, by Horn and Lin [29], Kowler and Kadlec [30], as well as Noorden et al. [31] for other applications. The method presented in this paper differs from the aforementioned ones in the means by which the periodic solutions of the state and costate equations are obtained. A pseudocode representation of the cyclic control optimization algorithm is shown in Algorithm 1.

Algorithm 1 Cyclic optimal control algorithm.

```
 1: //########### Basic functions of the control problem ###########
 2: define f(x, u){                          // RHS of state dynamics
 3:    return temporal rate of state vector;
 4: };
 5: define ζ(x, u, ā){                        // Path cost function, ā may contain W and Q_H
 6:    return path cost value;
 7: };
 8: define H(x, u, λ, ā){                     // Hamilton function
 9:    return ζ(x, u, ā) + λ^T f(x, u);
10: };
11: define -∇_x^app H(x, u, λ, ā){            // Approximation of RHS of costate dynamics
12:    declare ε = small number;
13:    declare ê_m = unit vector of m^th state component;
14:    return ∑_m -ê_m ( H(x + ε ê_m, u, λ, ā) − H(x, u, λ, ā) )/ ε;
15: };
16: define ∇_u^app H(x, u, λ, ā){             // Approximation of control gradient function
17:    declare ε = small number;
18:    declare ê_m = unit vector of m^th control component;
19:    return ∑_m ê_m ( H(x, u + ε ê_m, λ, ā) − H(x, u − ε ê_m, λ, ā) )/ (2 ε);
20: };
21: //########### Declare control, state, and costate functions ###########
22: declare τ = cycle time;                   // Define cycle time
23: declare u(t) with t ∈ [0, τ);             // Control vector function
24: declare x(t) with t ∈ [0, τ);             // State vector function
25: declare λ(t) with t ∈ [0, τ);             // Costate vector function
26: initialize u(t) with smooth guess of optimal control vector function;
27: initialize x(0) with arbitrary admissible values;
28: initialize λ(0) = 0;
29: //########### Iterative optimization of the control ###########
30: for(n = 0; n < large number; n = n + 1){
31:    while(ε > small number){                            // Solve state dynamics
32:       for(t = 0; t < τ; t += Δt){  x(t + Δt) = x(t) + f(x(t), u(t)) Δt;  }
33:       ε = ||x(τ) − x(0)||;
34:       x(0) = x(τ);
35:    }
36:    declare a(t) = some_function(x(t), u(t)) for t ∈ [0, τ);  // Auxiliary vector function
37:    declare ā = cycle average of u(t) for t ∈ [0, τ);        // Auxiliary vector cycle average
38:    while(ε > small number){                                  // Solve costate dynamics
39:       λ(τ) = λ(0);
40:       for(t = τ; t > 0; t −= Δt){  λ(t − Δt) = λ(t) − -∇_x^app H(x(t), u(t), λ(t), ā) Δt;  }
41:       ε = ||λ(τ) − λ(0)||;
42:    }
43:    declare s(t) = ∇_u^app H(x(t), u(t), λ(t)) for t ∈ [0, τ);  // Define search direction
44:    smooth s(t) in the periodic domain t ∈ [0, τ);             // Smooth search direction
45:    declare Λ = small number;                                  // Step-size factor
46:    update u(t) = u(t) + Λ s(t) for t ∈ [0, τ);                // Shift control
47:    smooth u(t) in the periodic domain t ∈ [0, τ);             // Smooth control
48: }
```

In the first section, between lines 1 and 20, the basic function definitions are made. First, the right-hand side (RHS) of the state dynamics is defined as a vector function $f(x, u)$. Afterwards, the path cost function $\zeta(x, u, \bar{a})$ that is to be maximized in terms of the objective functional from Equation (4) is defined. Here, the argument $\bar{a}$ may, for example, contain the cycle work W and heat Q_H, as required in Equation (15). Then, based on the definition of the Hamilton function $H(x, u, \lambda, \bar{a})$ from lines 8 to 10, the right-hand side of the costate dynamics $-\nabla_x^{\mathrm{app}} H(x, u, \lambda, \bar{a})$ is defined in terms of an approximation:

$$-\frac{\partial H(x, u, \lambda, \bar{a})}{\partial x_m} \approx -\frac{H(x + \varepsilon\,\hat{e}_m, u, \lambda, \bar{a}) - H(x, u, \lambda, \bar{a})}{\varepsilon} \tag{19}$$

where x_m is the mth state component, $\hat{e}_m$ is the corresponding unit vector and ε is a sufficiently small difference. For simple systems, it is reasonable to derive this right-hand side in terms of algebraic expressions. However, for involved systems it can be much more practical to do this by numerical partial differentiation as, for example, according to Equation (19). Note that for systems with discontinuous dynamics additional care must be taken here in order to make sure that the derivatives are properly approximated in the vicinity of the discontinuity. Similarly to $-\nabla_x^{\mathrm{app}} H(x, u, \lambda, \bar{a})$, the vector $\nabla_u^{\mathrm{app}} H(x, u, \lambda, \bar{a})$ of partial derivatives of the Hamilton function with respect to the control is defined from lines 16 to 20. Here, the following approximation is used:

$$\frac{\partial H(x, u, \lambda, \bar{a})}{\partial u_m} \approx \frac{H(x, u + \varepsilon\,\hat{e}_m, \lambda, \bar{a}) - H(x, u - \varepsilon\,\hat{e}_m, \lambda, \bar{a})}{2\varepsilon} \tag{20}$$

where u_m is the mth control component, $\hat{e}_m$ is the corresponding unit vector and ε is, again, a sufficiently small difference.

In the second section between lines 21 and 28, the most relevant variables are declared and initialized. These are the control vector function $u(t)$, the state vector function $x(t)$, and the costate vector function $\lambda(t)$. They are defined in the time domain $[0, \tau)$ and may be implemented as vectors of arrays, where the array lengths correspond to the number of time steps per period. That means that the values of u, x, and λ are saved at every time step in the periodic domain. The control vector function $u(t)$ is initialized with a smooth τ-periodic guess of the optimal control. Moreover, $x(0)$ is set to arbitrary but physically admissible values, whereas $\lambda(0)$ is set to **0**.

The iterative control optimization itself is described in the third section between lines 29 and 48. This starts at iteration $n = 0$ with the initial guess of the smooth τ-periodic control function $u^{(0)}(t)$, and the previously defined state and costate initial values $x^{(0)}(0)$ and $\lambda^{(0)}(0)$, respectively. Here, the bracketed superscript identifies the iteration n of the for-loop from lines 30 to 48. Since the control is predefined, the state and costate problem can be solved separately. The steps to obtain the subsequent iteration $u^{(n+1)}(t)$, according to Algorithm 1, can be summarized in simplified terms:

0.   If $n = 0$, set $x^{(0)}(0)$ to arbitrary physically admissible values and $\lambda^{(0)}(0) := \mathbf{0}$. Otherwise, set $x^{(n)}(0) := x^{(n-1)}(\tau)$ and $\lambda^{(n)}(0) := \lambda^{(n-1)}(\tau)$.

1.   For given $u^{(n)}(t)$: Solve

$$\dot{x}^{(n)}(t) = f(x^{(n)}(t), u^{(n)}(t))$$

by temporal forward integration in the periodic domain until the cyclic equilibrium is reached. Save auxiliary quantities, such as the cycle work W and the heat Q_H in $\bar{a}^{(n)}$. (Algorithm 1, lines 31 to 37: example with Euler method.)

2.   For given $u^{(n)}(t), x^{(n)}(t), \bar{a}^{(n)}$: Solve

$$\dot{\lambda}^{(n)}(t) = -\nabla_x H(x^{(n)}(t), u^{(n)}(t), \lambda^{(n)}(t), \bar{a}^{(n)})$$

by temporal backward integration in the periodic domain until the cyclic equilibrium is reached. (Algorithm 1, lines 38 to 42: example with Euler method.)

3. Calculate the search direction (See Algorithm 1, line 43.)

$$s^{(n)}(t) := \nabla_u H(x^{(n)}(t), u^{(n)}(t), \lambda^{(n)}(t)).$$

4. Calculate the next control vector function

$$u^{(n+1)}(t) := u^{(n)}(t) + \Lambda \, s^{(n)}(t)$$

with a sufficiently small positive step-size factor Λ. (Algorithm 1, line 46.)

These five steps are repeated for increasing n until $u^{(n)}(t)$ has converged to the optimal control. This can, for example, be checked with the below indicators which should both approach a numerical zero:

$$C_\nabla^{(n)} := \int_0^\tau \left(\nabla_u H(x^{(n)}(t), u^{(n)}(t), \lambda^{(n)}(t)) \right)^2 dt, \tag{21}$$

$$C_H^{(n)} := \max_t H(x^{(n)}(t), u^{(n)}(t), \lambda^{(n)}(t)) - \min_t H(x^{(n)}(t), u^{(n)}(t), \lambda^{(n)}(t)). \tag{22}$$

4.1. Exploiting Limit Cycles

The dynamic systems considered here are required to posses a limit cycle. This means that, given a proper periodic control function, the systems feature a closed state trajectory $x_\infty(t)$ in the state space, which is attractive. Here, the subscript ∞ refers to this limit cycle. If the state dynamics is integrated forward in time starting from $t = 0$ with a proper initial value $x|_0$ in the vicinity of $x_\infty(0)$, then $x|_t \to x_\infty(t)$ for $t \to \infty$. This is a property of many dissipative systems, and it is directly made use of in the first step (Algorithm 1, lines 31 to 35) for finding the periodic solution of the state vector.

Numerical studies of optimal control problems carried out in the context of this work show an interesting type of symmetry in that the costate dynamics then features a limit cycle, too, for prescribed $u(t)$ and $x|_t := x_\infty(t)$. However, the respective closed trajectory $\lambda_\infty(t)$ is repulsive when the costate dynamics is considered forward in time, and becomes attractive in the reversed time direction: $\lambda|_t \to \lambda_\infty(t)$ for $t \to -\infty$. Such a relation is not surprising, as the overall Hamiltonian system is conservative and the expansive costate dynamics compensates for the contractive state dynamics. Therefore, in the second step (Algorithm 1, line 38 to 42), the temporal direction of integration is reversed to obtain the periodic solution of the costate vector.

The correspondingly enhanced stability of the costate problem under temporal backward integration can also be exploited without directly making use of the limit-cycle property for obtaining the periodic solution, but for solving boundary value problems, for example, see [26,31].

4.2. Proper Embedding of the Control

Note that the way in which the control is embedded in the system's state dynamics, can influence the existence of these limit cycles in the state and costate problems. In the Stirling engine optimization problem, considered in this paper, our goal is to optimize the piston paths. Hence, we chose to structure the system dynamics in such a way that the control function determines the temporal evolution of the working space volumes V_i with $i \in \{H, L\}$. A rather obvious choice for embedding the control $u_i(t)$ would be:

$$\dot{V}_i = u_i(t). \tag{23}$$

For an explicitly time-dependent, τ-periodic control $u_i(t)$ that has an average value $\int_0^\tau u_i(t)\,\mathrm{d}t = 0$, the resulting volumes V_i will be τ-periodic. However, this dynamics does not feature a limit cycle: The solution of $V_i|_t$ will even for t being large integer multiples of τ remain equal to the initial value $V_i|_0$ from which integration was started. This will translate to the costate problem, so that the solution algorithm from above will, in general, not converge without further manipulation of the optimization problem.

A proper choice for embedding the control in the system's state dynamics that leads to the existence of a limit cycle for a continuous τ-periodic $u_i(t)$ is:

$$\dot{V_i} = \frac{\nu}{\tau}(u_i(t) - V_i) \tag{24}$$

with a sufficiently large number ν. This dynamics is similar to that of an overdamped mass-spring-damper system, where the spring support moves according to $u_i(t)$. If $\nu \to \infty$ then $V_i|_t \to u_i(t)$ for $t \to \infty$, independent of the initial value $V_i|_0$. That is, for $\nu \to \infty$, the control vector function $\boldsymbol{u}(t)$ represents the limit cycle $\boldsymbol{V}_\infty(t)$ of the volumes.

5. Results

The previously described optimization algorithm was applied to the exemplary alpha-Stirling engine model presented in Section 2 to optimize the control (piston paths) regarding both power and efficiency. The optimizations were performed for varying cycle time τ, or correspondingly varying engine speed $1/\tau$. With help of the penalty function from Equation (16), the working volumes were restricted to an approximate range from $100\,\mathrm{cm}^3$ to $1100\,\mathrm{cm}^3$. The initial control, from which the optimizations were started, corresponds to harmonic piston paths exploiting the complete admissible working volume range and featuring a phase shift of $90°$. This harmonic control will, in the following, also be used as a benchmark.

In Figure 2, the Stirling engine's power-optimal working volume trajectories are shown for different engine speeds.

It can be seen that the power-optimal volume trajectories significantly deviate from harmonic shapes. At low engine speeds, it is favorable to let the pistons rest in their extreme positions for about half the cycle time. This feature is less pronounced at higher engine speeds. Above ca. $1100\,\mathrm{rpm}$ and ca. $1300\,\mathrm{rpm}$ the power-optimal working volume trajectories detach from the maximum volume bounds, as can be seen in Figure 2 at $t/\tau \approx 1/4$ and $t/\tau \approx 1$, respectively. At the highest considered engine speed of $2000\,\mathrm{rpm}$, the swept volumes are considerably reduced. The power-optimal engine speed is ca. $920\,\mathrm{rpm}$, which is highlighted in Figure 2 by thick black lines. It is interesting that the power-optimal piston motions found in [22] by direct optimization are similar to the trajectories from Figure 2 for low and medium engine speeds. This is the case even though, in the current study, a Stirling engine with different parameters, and a much more detailed regenerator model, are considered. Nevertheless, compared to the parametric AS class of piston motions used in [22] for direct optimization, the trajectories from Figure 2 feature more details.

In Figure 3, the Stirling engine's efficiency-optimal working volume trajectories are shown, again for different engine speeds.

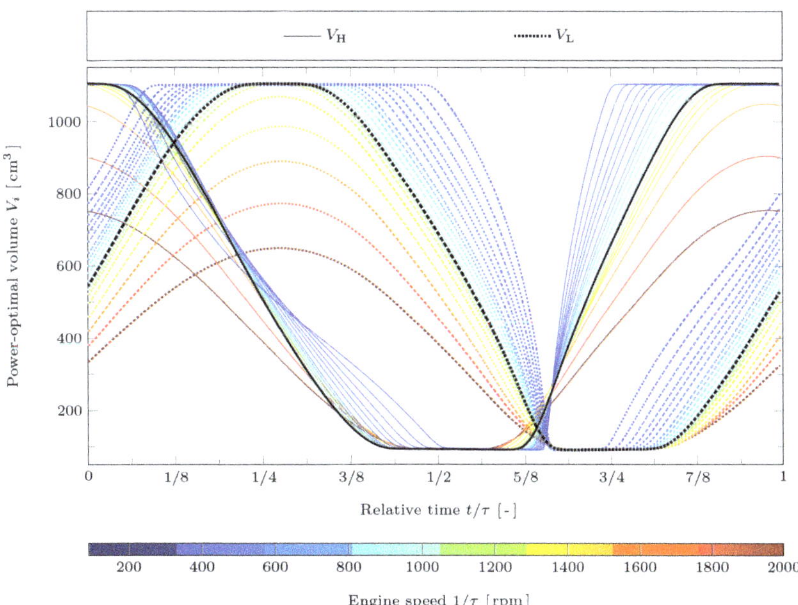

Figure 2. Power-optimal trajectories of the working volumes V_H (solid lines) and V_L (dotted lines) of an exemplary alpha-Stirling engine for varying engine speed $1/\tau$. The power-optimal engine speed is ca. 920 rpm (thick black lines).

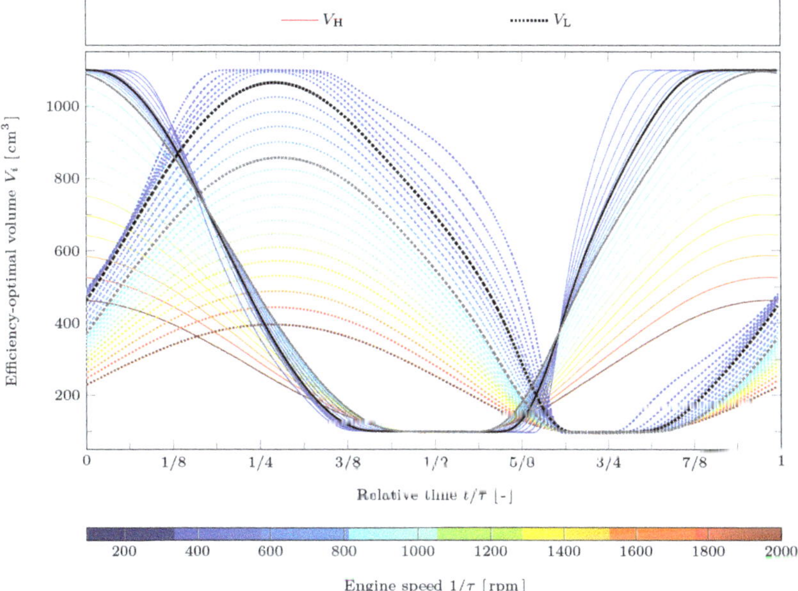

Figure 3. Efficiency-optimal trajectories of the working volumes V_H (solid lines) and V_L (dotted lines) of an exemplary alpha-Stirling engine for varying engine speed $1/\tau$. The efficiency-optimal engine speed is ca. 430 rpm (thick black lines). The power-optimal engine speed for the efficiency-optimal control is ca. 670 rpm (thick gray lines).

Obviously, the efficiency-optimal volume trajectories are different from the power-optimal ones. The tendency to let the pistons rest in their extreme positions at low engine speeds can also be observed here. However, especially for the cold working volume, this feature is less distinctive than with the power-optimal trajectories. The efficiency-optimal volume trajectories tend to involve smaller piston velocities and reduced swept volumes, when compared to the power-optimal trajectories for the same engine speed. The efficiency-optimal engine speed is ca. 430 rpm, which is highlighted in Figure 3 by thick black lines.

In Figure 4 the Stirling engine's power and efficiency are plotted against the engine speed for both the power-optimal control (solid, blue) and the efficiency-optimal control (dashed, green). Moreover, the values obtained with the harmonic control (dotted, grey) from which the optimizations were started is shown as a benchmark.

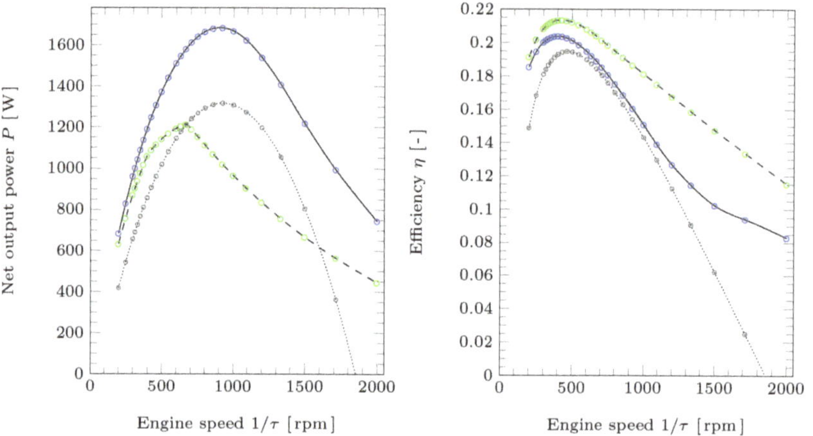

Figure 4. Net output power P and efficiency η for the harmonic (gray, dotted), power-optimal (blue, solid), and efficiency-optimal (green, dasehd) control plotted against the engine speed.

At the maximum power point, the relative power gain due to the power-optimal control is ca. 28%. The relative efficiency gain at the maximum efficiency point, which is achieved with the efficiency-optimal control, is ca. 10%.

Obviously, for a given engine speed, the power-optimal and efficiency-optimal controls each lead to maximum power and maximum efficiency, respectively. However, the optimization for one performance measure may come at the cost of the other. This can, in particular, be seen in the left-hand subfigure for the efficiency-optimal control: For medium engine speeds, the optimization for maximum efficiency leads to remarkable reductions in the output power. There, the power even drops below that obtained with the harmonic control. As opposed to that, the power-optimal control also leads to higher efficiency than the harmonic control over the complete considered range of engine speeds. Note, however, that this may be different for a different set of model parameters.

In the left-hand subfigure of Figure 4, it can be seen that the efficiency-optimal control (green, dashed) leads to a sharp bend at the maximum power point. This occurs at an engine speed of ca. 670 rpm, where the hot working volume trajectory (solid gray line in Figure 3) starts to detach from the maximum volume bound at $t/\tau \approx 1$.

Note that optimal controls obtained with the optimization algorithm introduced above generally constitute local optima. This becomes obvious in Figure 4 when, for example, comparing the powers resulting from the power-optimal controls at $\tau = 0.06\,\mathrm{s}$ (1000 rpm) and $\tau = 0.12\,\mathrm{s}$ (500 rpm). At $\tau = 0.06\,\mathrm{s}$ the power is close to the maximum value, whereas at $\tau = 0.12\,\mathrm{s}$ the power is much lower. However, it is clear that the power-optimal control obtained for $\tau = 0.06\,\mathrm{s}$ is an admissible periodic control for $\tau = 0.12\,\mathrm{s}$ as well, since two oscillations could be performed in one cycle. Hence, the power obtained for $\tau = 0.06\,\text{s}$ (1000 rpm) constitutes a lower bound for the global optimum at $\tau = 0.12\,\text{s}$ (500 rpm). It can be concluded that the optimal control found for $\tau = 0.12\,\text{s}$ (500 rpm) is a local optimum and not the global one. Similar arguments are applicable for the efficiency-optimal control, for example, with the efficiency values obtained at 400 rpm and 200 rpm.

6. Summary

In this paper, we presented an iterative gradient method for optimizing the piston paths of Stirling engines, which is based on Cyclic Optimal Control Theory. After a brief introduction to Stirling engines, we outlined an exemplary irreversible alpha-Stirling engine model. This model served for illustrative purposes and is structured in such a way that its state dynamics can be expressed as a set of coupled ordinary differential equations. The Stirling engine's state dynamics is influenced by an explicitly time-dependent, vector-valued control function. In particular, this control function determines the temporal paths of the engine's two working pistons.

The question of which realization of those piston paths—or correspondingly what control function—leads to optimal engine performance regarding a specific objective for a fixed cycle time, constitutes a cyclic optimal control problem. We introduced the necessary conditions of optimality for this problem, involving the definition of a Hamilton function, a path target function, and a set of adjoint ordinary differential equations (costate dynamics). For the optimization of Stirling engines, we presented proper definitions of path target functions to maximize either power or efficiency. Maximum and minimum volume constraints were accounted for here via an additional penalty term in the respective path target function.

Then, we gave a detailed description of an iterative optimization algorithm that starts with a predefined periodic control function, which is gradually shifted over the course of the iterations, so as to gradually enlarge the objective and approach the optimal control. To determine those gradual shifts of the control in every iteration, not only the engine's state dynamics needs to be solved, but also the costate dynamics. The cyclic optimal control problem features a symmetry that manifests itself in attractive and repulsive limit cycles in the state and costate dynamics, respectively. The algorithm exploits these limit cycles to solve the state and costate dynamics for periodic boundary conditions.

This optimization algorithm was applied to both, power-optimize and efficiency-optimize the control (piston paths) of the above-mentioned exemplary alpha-Stirling engine model for a range of cycle times. The optimization results were compared to piston paths that correspond to a harmonic control with $90°$ phase shift, which had been used as an initial control for the optimizations. At the maximum power point, the relative power gain due to the power-optimal control is ca. 28%, whereas the relative efficiency gain due to the efficiency-optimal control at the maximum efficiency point is ca. 10%.

The developed optimization algorithm can easily be applied to other Stirling engine models that involve additional irreversibilities or transfer laws that are adapted to describe a specific experimental engine setup. Design limitations such as maximum pressure or maximum piston acceleration can be included in the optimal control problem in terms of additional constraints. Moreover, other types of machines, such as beta-type or free-piston Stirling engines, Stirling cryocoolers, or Vuilleumier refrigerators can be optimized analogously.

7. Conclusions

For the cyclic optimal control problems of a Stirling engine investigated here, we observed that the existence of an attractive limit cycle in the state dynamics translates to a repulsive limit cycle in the costate dynamics. We conclude that a very stable and numerically efficient way to solve such problems is through using indirect iterative gradient algorithms that exploit these limit cycles to solve the state and costate dynamics by temporal forward and backward integration, respectively.

In the case of the considered exemplary Stirling engine, the conducted piston path optimizations lead to significant performance gains of ca. 28% in maximum power and ca. 10% in maximum efficiency. This generally confirms earlier results [22–24] and we conclude that it can be very worthwhile to perform such optimizations during the design process of new engines or to improve the control strategy of existing ones. Numerous types of energy conversion systems operate cyclically. We expect that cyclic optimal control theory and, in particular, the developed algorithm can be applied to a wide class of these systems, in order to identify potential performance improvements and enhance their sustainability.

Author Contributions: Investigation, R.P. and K.H.H. All authors have read and agreed to the published version of the manuscript.

Funding: This research was funded by the German Federal Ministry of Education and Research grant number FKZ01LY1706B.

Conflicts of Interest: The authors declare no conflict of interest.

References

1. Swedish Stirling AB. Residual Gases—A Huge Market Potential for PWR BLOK. Available online: https://swedishstirling.com/en/marknad/ (accessed on 19 January 2021).
2. Microgen Engine Corporation Group. Applications. Available online: https://www.microgen-engine.com/applications/ (accessed on 19 January 2021).
3. Azelio. Building a Renewable Future. Available online: https://www.azelio.com/product/ (accessed on 19 January 2021).
4. Qnergy. Remote Power. Available online: https://www.qnergy.com/remote-power/ (accessed on 19 January 2021).
5. Andresen, B.; Salamon, P.; Berry, R.S. Thermodynamics in Finite Time. *Phys. Today* **1984**, *37*, 62–70. [CrossRef]
6. Andresen, B. Current Trends in Finite-Time Thermodynamics. *Angew. Chem.* **2011**, *50*, 2690–2705. [CrossRef] [PubMed]
7. Rubin, M.H. Optimal Configuration of a Class of Irreversible Heat Engines. I. *Phys. Rev. A* **1979**, *19*, 1272–1276. [CrossRef]
8. Rubin, M.H. Optimal Configuration of a Class of Irreversible Heat Engines. II. *Phys. Rev. A* **1979**, *19*, 1277–1289. [CrossRef]
9. Hoffmann, K.H.; Burzler, J.M.; Schubert, S. Endoreversible Thermodynamics. *J. Non Equilib. Thermodyn.* **1997**, *22*, 311–355.
10. Hoffmann, K.H.; Burzler, J.M.; Fischer, A.; Schaller, M.; Schubert, S. Optimal Process Paths for Endoreversible Systems. *J. Non Equilib. Thermodyn.* **2003**, *28*, 233–268. [CrossRef]
11. Hoffmann, K.H.; Watowich, S.J.; Berry, R.S. Optimal Paths for Thermodynamic Systems: The Ideal Diesel Cycle. *J. Appl. Phys.* **1985**, *58*, 2125–2134. [CrossRef]
12. Burzler, J.M.; Blaudeck, P.; Hoffmann, K.H. Optimal Piston Paths for Diesel Engines. In *Thermodynamics of Energy Conversion and Transport*; Stanislaw Sieniutycz, S., de Vos, A., Eds.; Springer: New York, NY, USA, 2000; pp. 173–198. [CrossRef]
13. Chen, L.; Xia, S.; Sun, F. Optimizing piston velocity profile for maximum work output from a generalized radiative law Diesel engine. *Math. Comput. Model.* **2011**, *54*, 2051–2063. [CrossRef]
14. Xia, S.; Chen, L.; Sun, F. Engine performance improved by controlling piston motion: Linear phenomenological law system Diesel cycle. *Int. J. Therm. Sci.* **2012**, *51*, 163–174. [CrossRef]
15. Mozurkewich, M.; Berry, R.S. Optimal Paths for Thermodynamic Systems: The ideal Otto Cycle. *J. Appl. Phys.* **1982**, *53*, 34–42. [CrossRef]
16. Fischer, A.; Hoffmann, K.H. Can a quantitative simulation of an Otto engine be accurately rendered by a simple Novikov model with heat leak? *J. Non Equilib. Thermodyn.* **2004**, *29*, 9–28. [CrossRef]
17. Ge, Y.; Chen, L.; Sun, F. Optimal path of piston motion of irreversible Otto cycle for minimum entropy generation with radiative heat transfer law. *J. Energy Inst.* **2012**, *85*, 140–149. [CrossRef]
18. Watowich, S.J.; Hoffmann, K.H.; Berry, R.S. Intrinsically Irreversible Light-Driven Engine. *J. Appl. Phys.* **1985**, *58*, 2893–2901. [CrossRef]
19. Watowich, S.J.; Hoffmann, K.H.; Berry, R.S. Optimal Paths for a Bimolecular, Light-Driven Engine. *Il Nuovo Cim. B* **1989**, *104*, 131–147. [CrossRef]
20. Ma, K.; Chen, L.; Sun, F. Optimal paths for a light-driven engine with a linear phenomenological heat transfer law. *Sci. China Chem.* **2010**, *53*, 917–926. [CrossRef]
21. Kojima, S. Maximum Work of Free-Piston Stirling Engine Generators. *J. Non Equilib. Thermodyn.* **2017**, *42*, 169–186. [CrossRef]
22. Masser, R.; Khodja, A.; Scheunert, M.; Schwalbe, K.; Fischer, A.; Paul, R.; Hoffmann, K.H. Optimized Piston Motion for an Alpha-Type Stirling Engine. *Entropy* **2020**, *22*, 700. [CrossRef]
23. Scheunert, M.; Masser, R.; Khodja, A.; Paul, R.; Schwalbe, K.; Fischer, A.; Hoffmann, K.H. Power-Optimized Sinusoidal Piston Motion and Its Performance Gain for an Alpha-Type Stirling Engine with Limited Regeneration. *Energies* **2020**, *13*, 4564. [CrossRef]

24. Craun, M.; Bamieh, B. Optimal Periodic Control of an Ideal Stirling Engine Model. *J. Dyn. Syst. Meas. Control* **2015**, *137*, 071002. [CrossRef]
25. Paul, R.R. Optimal Control of Stirling Engines. Ph.D. Thesis, Technische Universität Chemnitz, Chemnitz, Germany, 2020.
26. Papageorgiou, M.; Leibold, M.; Buss, M. *Optimierung—Statische, Dynamische, Stochastische Verfahren für die Anwendung*; Springer Vieweg: Berlin/Heidelberg, Germany, 2015. [CrossRef]
27. Berry, R.S.; Kazakov, V.A.; Sieniutycz, S.; Szwast, Z.; Tsirlin, A.M. *Thermodynamic Optimization of Finite-Time Processes*; John Wiley & Sons: Chicester, UK, 2000.
28. Craun, M.J. Modeling and Control of an Actuated Stirling Engine. Ph.D. Thesis, University of California Santa Barbara, Santa Barbara, CA, USA, 2015.
29. Horn, F.J.M.; Lin, R.C. Periodic processes: A variational approach. *Ind. Eng. Chem. Process Des. Dev.* **1967**, *6*, 21–30. [CrossRef]
30. Kowler, D.E.; Kadlec, R.H. The optimal control of a periodic adsorber: Part II. Theory. *AIChE J.* **1972**, *18*, 1212–1219. [CrossRef]
31. van Noorden, T.L.; Verduyn Lunel, S.M.; Bliek, A. Optimization of cyclically operated reactors and separators. *Chem. Eng. Sci.* **2003**, *58*, 4115–4127. [CrossRef]

 *symmetry*

Review

Recent Advances on Boundary Conditions for Equations in Nonequilibrium Thermodynamics

Wen-An Yong [1,2,*] **and Yizhou Zhou** [3]

[1] Department of Mathematical Sciences, Tsinghua University, Beijing 100084, China
[2] Yanqi Lake Beijing Institute of Mathematical Sciences and Applications, Beijing 101408, China
[3] School of Mathematical Sciences, Peking University, Beijing 100871, China; zhouyz@math.pku.edu.cn
* Correspondence: wayong@tsinghua.edu.cn

Abstract: This paper is concerned with modeling nonequilibrium phenomena in spatial domains with boundaries. The resultant models consist of hyperbolic systems of first-order partial differential equations with boundary conditions (BCs). Taking a linearized moment closure system as an example, we show that the structural stability condition and the uniform Kreiss condition do not automatically guarantee the compatibility of the models with the corresponding classical models. This motivated the generalized Kreiss condition (GKC)—a strengthened version of the uniform Kreiss condition. Under the GKC and the structural stability condition, we show how to derive the reduced BCs for the equilibrium systems as the classical models. For linearized problems, the validity of the reduced BCs can be rigorously verified. Furthermore, we use a simple example to show how thus far developed theory can be used to construct proper BCs for equations modeling nonequilibrium phenomena in spatial domains with boundaries.

Keywords: hyperbolic relaxation system; structural stability condition; generalized Kreiss condition

Citation: Yong, W.-A.; Zhou, Y. Recent Advances on Boundary Conditions for Equations in Nonequilibrium Thermodynamics. *Symmetry* **2021**, *13*, 1710. https://doi.org/10.3390/sym13091710

Academic Editors: Róbert Kovács, Patrizia Rogolino, Francesco Oliveri, Charles F. Dunklm, Marin Marin and Mariano Torrisi

Received: 10 May 2021
Accepted: 1 September 2021
Published: 16 September 2021

Publisher's Note: MDPI stays neutral with regard to jurisdictional claims in published maps and institutional affiliations.

1. Introduction

Irreversible thermodynamics is a theory in physics for the mathematical modeling of nonequilibrium processes. The resultant models are usually time-dependent partial differential equations (PDEs) [1]. So far, the theory is still developing, and there are no well-accepted rules in establishing the equations. Therefore a number of different theories exist [2–8], leading to various PDEs. It is a challenging issue to evaluate the reasonableness of the different equations. To do so, four fundamental requirements were proposed and expounded in [9]. They are the observability of physical phenomena, time irreversibility, long-time tendency to equilibrium, and compatibility with possibly existing classical theories.

Recall well-known theories extended irreversible thermodynamics (EIT) [4,5]), rational extended thermodynamics (RET) [6]), and conservation–dissipation formalism (CDF) [8]). Irreversible processes can be described with PDEs of the form:

$$U_t + \sum_{j=1}^{d} F_j(U)_{x_j} = \tilde{Q}(U). \tag{1}$$

Here, $U = U(x,t) \in G \subset \mathbb{R}^n$ is the unknown function of $x = (x_1, x_2, , x_d)$ with d (=1, 2 or 3) for the spatial dimension and $t > 0$, G is an open set called the state space, $F_j(U)$ is the flux along the x_j-direction, source term $\tilde{Q}(U)$ represents the thermodynamical force, and subscripts t, x_j denote partial derivatives with respect to the corresponding variables. Very often, the source term can be written as

$$\tilde{Q}(U) = \begin{pmatrix} 0 \\ q(U) \end{pmatrix}.$$

Accordingly, we partition

$$U = \begin{pmatrix} u \\ v \end{pmatrix}, \quad F_j(U) = \begin{pmatrix} f_j(u,v) \\ g_j(u,v) \end{pmatrix}.$$

With this partition, $u \in \mathbb{R}^{n-r}$ stands for the conserved variables, and $v \in \mathbb{R}^r$ is referred to as nonequilibrium variables. To cover more equations from EIT [4,5] and general equation for nonequilibrium reversible–irreversible coupling (GENERIC) [7,10]), we also consider the following equation.

$$U_t + \sum_{j=1}^{d} A_j(U)U_{x_j} = \check{Q}(U), \tag{2}$$

which is more general than (1). Indeed, (1) can be written as (2) with $A_j(U) = \frac{\partial F_j}{\partial U}(U)$.

For such first-order PDEs, the first fundamental requirement corresponds to the hyperbolicity of Equation (2) [11,12]. To see the other requirements, thermodynamic force $Q(U)$ contains a relaxation time ϵ in general, and we explicitly write it as $\check{Q}(U) = Q(U)/\epsilon$. The requirements are closely related to the limit as the relaxation time ϵ tends to zero— the so-called zero relaxation limit. Without considering the boundary conditions, the four requirements are fulfilled by (2) if the PDEs satisfy the structural stability condition proposed in [13,14]. See [14,15].

When the nonequilibrium phenomena occur in a spatial domain with boundaries, the PDEs alone cannot completely describe the physical processes, and proper boundary conditions (BCs) are indispensable. However, it is a challenging task to have such BCs, since the physical meaning of the nonequilibrium variables is often unclear (e.g., the higher-order moments in moment-closure systems [6,16]). On the other hand, formulating proper BCs for hyperbolic systems of PDEs is mathematically a tough issue [17]. For example, the number of BCs for (2) should be equal to the number of positive eigenvalues of coefficient matrix $A_1(U)$ in case that the boundary is $x_1 = 0$; moreover, the BCs should satisfy the uniform Kreiss condition (UKC) [18] to ensure the well-posedness of the complete model (PDEs together with BCs)—the first fundamental requirement. Furthermore, the rest requirements also apply to the complete model. This suggests to study the zero relaxation limit of initial-boundary-value problems (IBVPs) for hyperbolic systems (1) or (2). The resultant results are expected to be useful in providing reasonable constraints for the possible BCs and in developing a systematical method to construct the required BCs.

The paper presents the first author's considerations in the past two decades and recent results on these issues. It is organized as follows. Section 2 contains some basic knowledges about relaxation system (3) and boundary conditions. In Section 3, we show that the structural stability condition and the UKC do not automatically guarantee compatibility, and present a strengthened version of the UKC—the generalized Kreiss condition (GKC). Section 4 is devoted to the derivation of the boundary condition for the zero relaxation limit under the GKC and the structural stability condition. In Section 5, we use an example to show how the theory presented above can be used to construct BCs for the PDEs (3). Some concluding remarks are given in Section 6.

2. Preliminaries

To address the long-term tendency and compatibility requirements, we explicitly introduce relaxation time ϵ and write Equation (3) as

$$U_t + \sum_{j=1}^{d} A_j(U)U_{x_j} = \frac{1}{\epsilon}Q(U). \tag{3}$$

For such a small parameter problem, the most fundamental is the following structural stability condition proposed in [13,14]:

(i)   There is an invertible $n \times n$-matrix $P(U)$ and an invertible $r \times r$-matrix $S(U)$ $(0 < r \leq n)$, defined on equilibrium manifold $\mathcal{E} = \{U \in G : Q(U) = 0\}$, such that

$$P(U)Q_U(U) = \begin{pmatrix} 0 & 0 \\ 0 & S(U) \end{pmatrix} P(U) \quad \text{for } U \in \mathcal{E};$$

(ii)   As a hyperbolic system, (1) is symmetrizable, that is, there is a positive definite symmetric matrix $A_0(U)$, such that

$$A_0(U)A_j(U) = A_j^*(U)A_0(U) \quad \text{for } U \in G;$$

(iii) The hyperbolic part and the source term are coupled in the following sense.

$$A_0(U)Q_U(U) + Q_U^*(U)A_0(U) \leq -P^*(U)\begin{pmatrix} 0 & 0 \\ 0 & I_r \end{pmatrix}P(U) \quad \text{for } U \in \mathcal{E}.$$

Here and below, Q_U denotes the Jacobian matrix of $Q = Q(U)$, I_k denotes the unit matrix of order k, and superscript $*$ denotes the (conjugate) transpose of a matrix.

In the structural stability condition, (i) is just the usual assumption in the corresponding theory for ordinary differential equations [14], that is, only for the source term; (ii) is for the part of spatial derivatives and means that (3) is symmetrizable hyperbolic; and (iii) is a coupling condition that involves the both parts. As observed in [14,19], the structural stability condition is fulfilled by many classical models from mathematical physics. It is related to the Onsager reciprocal relation and can be viewed as a stability criterion for nonequilibrium thermodynamics. Regarding the Onsager relation, matrix $A_0(U)Q_U(U)$ is often symmetric. For simplicity, this symmetry is assumed throughout the paper.

Under the structural stability condition, it was proved in [13,14] that the solution to the initial-value problem of (3) converges to that of the corresponding equilibrium system as ϵ goes to zero. For the case where

$$Q(U) = \begin{pmatrix} 0 \\ q(U) \end{pmatrix}$$

with $q = q(U) \in \mathbb{R}^r$, we write

$$U = \begin{pmatrix} u \\ v \end{pmatrix}, \qquad A_j(U) = \begin{pmatrix} A_{j11} & A_{j12} \\ A_{j21} & A_{j22} \end{pmatrix}(U).$$

In this case, it is reasonable to assume that $q(U) = 0$ if and only if $v = h(u)$. Then, the equilibrium system can be expressed as

$$v = h(u),$$

$$u_t + \sum_{j=1}^{d} A_{j11}(u, h(u))u_{x_j} + \sum_{j=1}^{d} A_{j12}(u, h(u))h(u)_{x_j} = 0.$$

Under the structural stability condition, it was also proved in [13,20] that the equilibrium system is symmetrizable hyperbolic.

When Equation (3) is given in a spatial domain with boundaries, proper boundary conditions (BCs) should be prescribed to solve the equations. By referring to [21], it is adequate to consider that the spatial domain is half-space

$$(x_1, x_2, \ldots, x_d) \equiv (x_1, \hat{x}) \in [0, +\infty) \times \mathbb{R}^{d-1}.$$

For this domain, assume that coefficient matrix $A_1(U)$ has p positive eigenvalues. According to the classical theory of hyperbolic equations [17,22], p independent BCs

$$B(U(0, \hat{x}, t); t) = 0 \tag{4}$$

should be prescribed at boundary $x_1 = 0$. Moreover, they should satisfy the uniform Kreiss condition (UKC) [18] to ensure the well-posedness, namely, there is a positive constant c_K, such that

$$\left| \det\left\{ B_U R_{\tilde{M}}^S(\xi, \omega) \right\} \right| \geq c_K \sqrt{\det\{ R_{\tilde{M}}^{S*}(\xi, \omega) R_{\tilde{M}}^S(\xi, \omega) \}}$$

for all $\omega \in \mathbb{R}^{d-1}$ and all complex number ξ with $Re\xi > 0$. To decode this UKC, we assume that $A_1(U)$ is invertible, and know from [18] that matrix

$$\tilde{M}(\xi, \omega) := A_1^{-1}\left(-\xi I + i \sum_{j=2}^{d} \omega_j A_j \right),$$

has p stable eigenvalues (eigenvalues with negative real parts) for $Re\xi > 0$ and $\omega = (\omega_2, \ldots, \omega_d) \in \mathbb{R}^{d-1}$. Then, there is a full-rank $n \times p$-matrix $R_{\tilde{M}}^S(\xi, \omega)$ (so-called right-stable matrix of $\tilde{M}$ [23]) satisfying

$$\tilde{M}(\xi, \omega) R_{\tilde{M}}^S(\xi, \omega) = R_{\tilde{M}}^S(\xi, \omega) \tilde{\Lambda}(\xi, \omega)$$

with $\tilde{\Lambda}(\xi, \omega)$ a $p \times p$-matrix having p stable eigenvalues.

When $A_1(U)$ is not invertible, boundary $x_1 = 0$ is referred to as characteristic. In this case, a modified UKC was formulated in [21] as a sufficient and essentially necessary condition for well-posedness. In addition, it was indicated in [21] that the boundary condition should not involve characteristic modes corresponding to the zero eigenvalue.

3. Generalized Kreiss Condition

This section shows that the structural stability condition together with the UKC do not automatically guarantee a well-behaved zero relaxation limit. To do this, we start with the linearized version of PDEs in (3):

$$U_t + \sum_{j=1}^{d} A_j U_{x_j} = \frac{1}{\epsilon} QU, \tag{5}$$

together with homogeneous boundary conditions

$$BU(0, \hat{x}, t) = 0 \tag{6}$$

for $x = (x_1, \hat{x}) = (x_1, x_2, \ldots, x_d)$ with $x_1 > 0$. Here, $A_j(j = 1, 2, \ldots, d)$ and Q are $n \times n$ constant matrices and B is a full-rank constant $p \times n$-matrix with p the number of positive eigenvalues of A_1.

Following [18,23,24], we consider the corresponding eigenvalue problem:

$$\begin{cases} \xi \hat{U} + A_1 \hat{U}_{x_1} - i \sum_{j=2}^{d} \omega_j A_j \hat{U} = Q\hat{U}, \\ B\hat{U}(0) = 0 \end{cases} \tag{7}$$

for complex number ξ with $Re\xi > 0$ and $\omega = (\omega_2, \ldots, \omega_d) \in \mathbb{R}^{d-1}$. Let $\hat{U} = \hat{U}(x_1)$ be a bounded solution to the above problem. It is not difficult to verify that

$$U^{\epsilon}(x, t) := \exp\left(\frac{\xi t}{\epsilon} - \frac{i\omega \cdot \hat{x}}{\epsilon} \right) \hat{U}\left(\frac{x_1}{\epsilon} \right)$$

solves (5) and (6) with a bounded initial value. Since $Re\xi > 0$, such a solution exponentially increases as ϵ goes to zero for any $t > 0$. Apparently, the zero relaxation limit is not well-behaved or does not exist in this situation.

In order to see when problem (7) has bounded solutions, we assume A_1 to be invertible and rewrite (7) as

$$\frac{d\hat{U}}{dx_1} = M(\xi, \omega, 1)\hat{U} \tag{8}$$

with

$$M(\xi, \omega, \eta) = A_1^{-1}\left(\eta Q - \xi I + i \sum_{j=2}^{d} \omega_j A_j\right).$$

Here, $\eta \geq 0$ was introduced to cover the case without source term Q, namely, $M(\xi, \omega, 0) = \tilde{M}(\xi, \omega)$ in Section 2.

According to Lemma 2.3 in [23], under the structural stability condition, matrix $M = M(\xi, \omega, \eta)$ has precisely p stable eigenvalues (with negative real parts) and $(n - p)$ unstable eigenvalues. Then, there is a full-rank $n \times p$-matrix $R_M^S = R_M^S(\xi, \omega, \eta)$ satisfying

$$MR_M^S = R_M^S\Lambda$$

with Λ a $p \times p$-matrix having p stable eigenvalues. If the $p \times p$-matrix $BR_M^S(\xi, \omega, 1)$ is not invertible, then there is a nonzero p-vector ζ, such that $BR_M^S(\xi, \omega, 1)\zeta = 0$. With $\hat{U}(0) = R_M^S(\xi, \omega, 1)\zeta$ as initial value, the ordinary differential Equation (8) has a bounded solution for $x_1 > 0$.

Consequently, we obtain

Proposition 1. *If there exists ξ with $Re\xi > 0$ such that the $p \times p$-matrix $BR_M^S(\xi, \omega, 1)$ is singular, then the problem in (5) and (6), with a bounded initial value, admits an exponentially increasing solution for $t > 0$ as ϵ goes to zero.*

The above discussion can be illustrated with the one-dimensional linearized moment closure system [16,25]:

$$\begin{cases} U_t + A_1 U_x = QU/\epsilon, \\ BU(0, t) = 0 \end{cases} \tag{9}$$

with $Q = \mathrm{diag}(0, 0, 0, -1)$ and

$$A_1 = \begin{pmatrix} 0 & \rho_0 & 0 & 0 \\ \theta_0/\rho_0 & 0 & 1 & 0 \\ 0 & 2\theta_0 & 0 & 6/\rho_0 \\ 0 & 0 & \rho_0\theta_0/2 & 0 \end{pmatrix}, \quad B = \begin{pmatrix} 0 & 1 & 0 & 0 \\ 0 & 0 & \chi\rho_0\sqrt{\theta_0} & -1 \end{pmatrix}.$$

Here, $U = (\rho, u, \theta, f)$ is the unknown. ρ, u, θ are fluid density, velocity, and temperature, respectively, while f is related to the heat flux. Constants ρ_0 and θ_0 in coefficient matrix A_1 denote density and temperature at the equilibrium. Constant χ in the BC is a positive parameter. Obviously, the moment-closure system satisfies the structural stability condition with symmetrizer

$$A_0 = \begin{pmatrix} 2\theta_0^3 & 0 & 0 & 0 \\ 0 & 2\rho_0^2\theta_0^2 & 0 & 0 \\ 0 & 0 & \rho_0^2\theta_0 & 0 \\ 0 & 0 & 0 & 12 \end{pmatrix}.$$

It was shown in [25] that the BC in (9) satisfies the UKC if and only if

$$\chi \neq \frac{\sqrt{3+\sqrt{6}}+\sqrt{3-\sqrt{6}}}{2(3+\sqrt{3})}.$$

Moreover, Proposition 3.3 in [25] says that if

$$\chi \in (0, \frac{\sqrt{3+\sqrt{6}}+\sqrt{3-\sqrt{6}}}{2(3+\sqrt{3})}),$$

there is a positive ζ such that $BR^S_M(1,\zeta)$ is singular. According to Proposition 1, the structural stability condition and the UKC do not automatically imply the existence of the zero relaxation limit.

Proposition 1 and the example above indicate the following necessary condition for the existence of zero relaxation limit:

$$\det\{BR^S_M(\xi,\omega,1)\} \neq 0, \quad \text{for any} \quad Re\xi > 0, \omega \in \mathbb{R}^{d-1}.$$

Motivated by this, the following condition was proposed in [23]:

Generalized Kreiss Condition. There exists a constant $c_K > 0$, such that

$$\left| \det\left\{ BR^S_M(\xi,\omega,\eta) \right\} \right| \geq c_K \sqrt{\det\{R^{S*}_M(\xi,\omega,\eta)R^S_M(\xi,\omega,\eta)\}}$$

for all $\eta \geq 0, \omega \in \mathbb{R}^{d-1}$ and ζ with $Re\xi > 0$.

Clearly, this GKC does not depend on the special choice of $R^S_M(\xi,\omega,\eta)$. It recovers the standard UKC when $\eta = 0$. Unlike the UKC, the GKC involves parameters even for one-dimensional problems. Moreover, it was shown in [23] that the GKC holds if the BCs are strictly dissipative in the following sense:

Definition 1. *The boundary condition is referred to as strictly dissipative with respect to symmetrizer A_0 and A_1, if there is a positive constant c_d, such that*

$$y^* A_0 A_1 y \leq -c_d |y|^2 + c_d^{-1} |By|^2$$

for all $y \in \mathbb{R}^n$.

We conclude this section with the following remark:

Remark 1. *In studying IBVPs for hyperbolic systems, it is important to distinguish whether the boundary is characteristic or not [21], that is, whether A_1 is invertible or not. In the above discussions, the invertibility of A_1 plays the role of the starting point. When A_1 is not invertible, a modified GKC was proposed in [26]. It covers the above GKC for noncharacteristic problems.*

4. Reduced Boundary Conditions

Besides the GKC, deriving reduced boundary conditions is another crucial issue in studying the zero relaxation limit for the initial-boundary-value problems. The relaxation limit satisfies the equilibrium system under the structural stability condition [14]. When the problem is given in the half-space, the equilibrium system alone can not determine the limit. It must be supplemented with proper boundary conditions. As part of the system determining the limit, such BCs should be completely derived from the relaxation system and its BCs. BCs thus derived are called reduced boundary conditions.

For linearized system (3) satisfying the structural stability condition, we assume without loss of generality that $Q = \mathrm{diag}(0, S)$ with S being a negative definite matrix. Corresponding to the partition of Q, we write coefficient matrix A_j as

$$A_j = \begin{pmatrix} A_{j11} & A_{j12} \\ A_{j21} & A_{j22} \end{pmatrix}.$$

Thus, the aforementioned equilibrium system reads as

$$\begin{cases} v = 0, \\ u_t + \sum_{j=1}^{d} A_{j11} u_{x_j} = 0. \end{cases} \tag{10}$$

Recall that coefficient matrices A_{j11} are symmetrizable.

Let p_1 be the number of positive eigenvalues for matrix A_{111}. According to the classical theory of IBVPs for hyperbolic systems [17,22], equilibrium system (10) needs p_1 BCs satisfying the UKC to be well-posed. As mentioned before, these BCs should be completely determined from the relaxation system and its BCs:

$$BU(0, \hat{x}, t) = (B_u, B_v) \begin{pmatrix} u \\ v \end{pmatrix} (0, \hat{x}, t) = b(t). \tag{11}$$

In this regard, it was established in [23,27] that

Theorem 1. *Under the structural stability condition and the GKC, there exists a $p_1 \times p$-matrix B_p, unique up to an invertible $p_1 \times p_1$-matrix multiplying B_p from left, such that relation*

$$B_p B_u u(0, \hat{x}, t) = B_p b(\hat{x}, t) \tag{12}$$

satisfies the UKC as a BC for the equilibrium system. Moreover, if boundary $x_1 = 0$ is characteristic for the equilibrium system, BC (12) does not involve the characteristic modes corresponding to the zero eigenvalue.

The equilibrium system with the reduced BC (12) consist of a well-posed IBVP since the reduced BC satisfies the UKC. The proof of this theorem involves the perturbation theory of linear operators [28] and subtle matrix analysis. The key is to analyze the limit of right-stable matrix $R_M^S(\xi, \omega, \eta)$ as $\eta \to \infty$. According to the GKC, $BR_M^S(\xi, \omega, +\infty)$ is a $p \times p$ invertible matrix. Then, B_p can be chosen as the $p_1 \times p$-matrix consisting of the first p_1 rows of $[BR_M^S(\xi, \omega, +\infty)]^{-1}$. Such a matrix B_p is independent of ξ and ω. We omit the details here, and the interested reader is referred to to [23,27].

Regarding this theorem:

Remark 2.

- *For nonlinear problems, reduced BCs are quite complicated, but can be derived by considering boundary-layer equations*

$$A_1(U)U_y = Q(U), \qquad y = \frac{x_1}{\epsilon} \geq 0.$$

For the details, the interested reader is referred to [23].
- *In case the boundary is characteristic (for the relaxation system), a similar result can be found in [26].*

To see the validity of the reduced BCs, we consider asymptotic solutions of the form:

$$\begin{pmatrix} u_\epsilon \\ v_\epsilon \end{pmatrix}(x_1, \hat{x}, t; \epsilon) = \begin{pmatrix} \bar{u} \\ \bar{v} \end{pmatrix}(x_1, \hat{x}, t) + \begin{pmatrix} \mu_0 \\ \nu_0 \end{pmatrix}(\frac{x_1}{\epsilon}, \hat{x}, t)$$

$$+ \begin{pmatrix} \mu_1 \\ \nu_1 \end{pmatrix}(\frac{x_1}{\sqrt{\epsilon}}, \hat{x}, t) + \sqrt{\epsilon}\begin{pmatrix} \mu_2 \\ \nu_2 \end{pmatrix}(\frac{x_1}{\sqrt{\epsilon}}, \hat{x}, t) \tag{13}$$

for relaxation system (5) and its BC (11). Here, $(\bar{u}, \bar{v})$ is the outer solution, and (μ_i, ν_i) $(i = 0, 1, 2)$ are boundary-layer corrections satisfying matching conditions

$$(\mu_i, \nu_i)(\infty, \hat{x}, t) = 0, \qquad i = 0, 1, 2.$$

By a standard procedure [27], the outer solution solves equilibrium system (10) with reduced BC (12), while the other terms in (13) can also be determined.

If the boundary is not characteristic for the equilibrium system, there is no boundary-layer with length $\sqrt{\epsilon}$. In other words, the last two terms in (13) are void, and the three scales in the expansion degenerate to two scales.

The validity of the reduced BC (12) is verified with the following theorem.

Theorem 2 ([27]). *Assuming that the structural stability condition and the GKC hold, the initial data are given at the equilibrium, and the BC is compatible with the initial condition, there exists a constant $K > 0$, such that the exact solution (u^ϵ, v^ϵ) fulfils the following estimate.*

$$\|(u^\epsilon - u_e, v^\epsilon - v_e)(\cdot, \cdot, t)\|_{L^2(\mathbb{R}^+ \times \mathbb{R}^{d-1})} \le K\epsilon^{1/2}$$

for all time $t \in [0, T]$.

Next, we return to one-dimensional linearized moment closure system (9) to illustrate the reduced BCs. For this model, we have

$$A_1 = \begin{pmatrix} A_{11} & A_{12} \\ A_{21} & A_{22} \end{pmatrix} = \left(\begin{array}{ccc|c} 0 & \rho_0 & 0 & 0 \\ \theta_0/\rho_0 & 0 & 1 & 0 \\ 0 & 2\theta_0 & 0 & 6/\rho_0 \\ \hline 0 & 0 & \rho_0\theta_0/2 & 0 \end{array}\right),$$

and

$$B = (B_u, B_v) = \left(\begin{array}{ccc|c} 0 & 1 & 0 & 0 \\ 0 & 0 & \chi\rho_0\sqrt{\theta_0} & -1 \end{array}\right), \qquad b(t) = 0.$$

As shown in [25], coefficient matrix A_1 has two positive eigenvalues, $\sqrt{3 \pm \sqrt{6}}\sqrt{\theta_0}$, and A_{11} has one positive eigenvalue $\sqrt{3\theta_0}$, that is, $p = 2$ and $p_1 = 1$. Thus equilibrium system

$$\begin{pmatrix} \bar{\rho} \\ \bar{u} \\ \bar{\theta} \end{pmatrix}_t + \begin{pmatrix} 0 & \rho_0 & 0 \\ \theta_0/\rho_0 & 0 & 1 \\ 0 & 2\theta_0 & 0 \end{pmatrix}\begin{pmatrix} \bar{\rho} \\ \bar{u} \\ \bar{\theta} \end{pmatrix}_{x_1} = 0$$

needs one reduced boundary condition. On the other hand, in this case, we have

$$BR_M^S(\xi, +\infty) = \begin{pmatrix} \sqrt{3\theta_0}/\rho_0 & 0 \\ 2\chi\theta_0\sqrt{\theta_0} & -\chi\theta_0\sqrt{\theta_0} \end{pmatrix}.$$

This matrix is obviously invertible, and its inverse is a lower-triangular matrix. Therefore, B_p can be chosen as the 1×2-matrix $(1, 0)$, and the reduced BC is

$$B_p B \bar{U}(0, t) = (1, 0) \begin{pmatrix} 0 & 1 & 0 & 0 \\ 0 & 0 & \chi \rho_0 \sqrt{\theta_0} & -1 \end{pmatrix} \bar{U}(0, t) = B_p b(t) = 0$$

with $\bar{U} = (\bar{\rho}, \bar{u}, \bar{\theta}, \bar{f})$. That is, just no-slip velocity BC

$$\bar{u}(0, t) = 0.$$

5. Construction of BCs

In this section, we show how the theory presented above can be used to construct BCs for PDEs (2). Usually, the relaxation system is derived by augmenting certain classical equations to take account into more refined physical processes. However, proper BCs for the relaxation systems are not always available, since the physical meaning of the nonequilibrium variables is often unclear (e.g., the higher-order moments in moment-closure systems [6,16]). Because of this, no physical means can be expected to obtain the BCs for certain relaxation systems.

For our construction of the BCs, we assume that both the relaxation systems and the well-posed BCs for the corresponding equilibrium systems are given. This assumption is reasonable because there already exist many well-known approaches to construct the relaxation systems, and much knowledge on equilibrium systems and their BCs exists.

We illustrate the construction with the following simple model.

$$\partial_t u - \partial_x v = 0,$$

$$\partial_t v - \partial_x \sigma = 0, \tag{14}$$

$$\partial_t(\sigma - Eu) = (g(u) - \sigma)/\epsilon.$$

This model was proposed in [29] for the isothermal motions of a viscoelastic material, and can be viewed as an augmentation of classical equations of isothermal elastodynamics:

$$\partial_t u - \partial_x v = 0,$$

$$\partial_t v - \partial_x g(u) = 0. \tag{15}$$

Here, u and σ denote strain and stress, v is related to particle velocity, E is a positive constant called dynamic Young's modulus, ϵ is the relaxation time, and $g(u)$ is a given function of u. For this model, the structural stability condition reads as

$$0 < g'(u) < E.$$

For simplicity, we assume that the model was defined in the domain where $x \geq -\alpha t$ and $g(u) = Gu$, with $G > 0$ a constant and α a parameter. Under change in variables $(x, t) \to (x + \alpha t, t)$, the system (14) becomes

$$\begin{pmatrix} u \\ v \\ p \end{pmatrix}_t + \begin{pmatrix} \alpha & -1 & 0 \\ -G & \alpha & -1 \\ 0 & G - E & \alpha \end{pmatrix} \begin{pmatrix} u \\ v \\ p \end{pmatrix}_x = \frac{1}{\epsilon} \begin{pmatrix} 0 & 0 & 0 \\ 0 & 0 & 0 \\ 0 & 0 & -1 \end{pmatrix} \begin{pmatrix} u \\ v \\ p \end{pmatrix} \tag{16}$$

with $p = \sigma - Gu$ and (15) reads as

$$\begin{pmatrix} u \\ v \end{pmatrix}_t + \begin{pmatrix} \alpha & -1 \\ -G & \alpha \end{pmatrix} \begin{pmatrix} u \\ v \end{pmatrix}_x = 0. \tag{17}$$

Moreover, the given BC for the equilibrium system is

$$\hat{B}\left(\begin{array}{c} u \\ v \end{array}\right)(0,t) = \hat{b}(t)$$

with $\hat{B} = (\hat{B}_1, \hat{B}_2)$ being a constant full-rank matrix. The number of rows of $\hat{B}$ is equal to the number of positive eigenvalues of coefficient matrix

$$A_{11} \equiv \left(\begin{array}{cc} \alpha & -1 \\ -G & \alpha \end{array}\right).$$

This matrix has eigenvalues $\alpha \pm \sqrt{G}$, while coefficient matrix

$$A_1 \equiv \left(\begin{array}{ccc} \alpha & -1 & 0 \\ -G & \alpha & -1 \\ 0 & G-E & \alpha \end{array}\right)$$

has eigenvalues $\alpha \pm \sqrt{E}$ and α.

To be precise and simple, we chose $\alpha \in (-\sqrt{E}, -\sqrt{G})$, while other choices can be found in [30]. For such an α, boundary $x = 0$ is noncharacteristic for both (17) and (16), boundary matrix $\hat{B}$ is void, and we need to construct one BC of the form

$$B\left(\begin{array}{c} u(0,t) \\ v(0,t) \\ p(0,t) \end{array}\right) = b_\epsilon(t) \tag{18}$$

for the relaxation system (16).

To find such a nonzero vector $B \in \mathbb{R}^3$, we consider the asymptotic solution of form

$$\left(\begin{array}{c} u_\epsilon \\ v_\epsilon \\ p_\epsilon \end{array}\right)(x,t;\epsilon) = \left(\begin{array}{c} \bar{u} \\ \bar{v} \\ \bar{p} \end{array}\right)(x,t) + \left(\begin{array}{c} \mu \\ \nu \\ \pi \end{array}\right)(\frac{x}{\epsilon},t)$$

and require it to satisfy the BC (18):

$$B\left(\begin{array}{c} \bar{u} + \mu \\ \bar{v} + \nu \\ \bar{p} + \pi \end{array}\right)(0,t) = b_0(t).$$

In this way, we obtain some algebraic relations or restrictions on B, from which we deduce that

$$B^* = \left(\begin{array}{c} 1 + \dfrac{C_1}{\alpha^2 - G} \\[2mm] \sqrt{G} + \dfrac{\alpha C_1}{\alpha^2 - G} \\[2mm] C_1 \end{array}\right) \times \left(\begin{array}{c} 1 + \dfrac{C_2}{\alpha^2 - G} \\[2mm] -\sqrt{G} + \dfrac{\alpha C_2}{\alpha^2 - G} \\[2mm] C_2 \end{array}\right)$$

with C_1, C_2 two free parameters. BCs thus constructed are not unique, which is expected.

It was shown in [30] that, when C_1, C_2 are both sufficiently small, the constructed BCs are strictly dissipative. Thus, they satisfy the GKC. On this basis, the validity of the constructed BCs was rigorously proved in [30].

6. Concluding Remarks

In this paper, we presented a systematical review on boundary conditions (BCs) for partial differential equations (PDEs) from nonequilibrium thermodynamics. From a stability point of view, such PDEs should satisfy the structural stability condition. In particular, they constitute hyperbolic systems, for which the uniform Kreiss condition (UKC) is a sufficient and essentially necessary condition for the well-posedness of the corresponding models (PDEs with BCs). Taking a linearized moment closure system as an example, we showed that the structural stability condition and UKC do not automatically guarantee the compatibility of the models with the corresponding classical models. This motivated the generalized Kreiss condition (GKC)—a strengthened version of the UKC. It is expected that the GKC and the structural stability condition are the very criteria to verify the four fundamental requirements when modeling nonequilibrium phenomena in spatial domains with boundaries.

The compatibility and long-time tendency requirements suggest to study the zero relaxation limit of the PDEs with BCs. It is trivial to obtain the equilibrium system for the limit. However, the equilibrium system alone cannot determine the limit, and proper BCs are needed. The needed BCs are referred to as reduced BCs, and they should be completely determined by the PDEs with the given BCs. The derivation of the reduced BCs is by no means trivial.

Under the GKC and the structural stability condition, we showed how to derive the reduced BCs for general relaxation models. For linearized problems, the validity of the reduced BCs was rigorously verified. For nonlinear problems, such a verification is open. Furthermore, we used a simple example to show how the thus far developed theory can be used to construct proper BCs accompanying PDEs modeling nonequilibrium phenomena in spatial domains with boundaries.

Author Contributions: Conceptualization, W.-A.Y.; investigation, W.-A.Y., Y.Z.; writing—original draft preparation, Y.Z., writing—review and editing, W.-A.Y., Y.Z.; project administration, W.-A.Y.; funding acquisition, W.-A.Y. All authors have read and agreed to the published version of the manuscript.

Funding: This research was funded by the National Natural Science Foundation of China (grant number 12071246).

Conflicts of Interest: The authors declare no conflict of interest.

References

1. Van, P. Nonequilibrium thermodynamics: Emergent and fundamental. *Philos. Trans. R. Soc. A* **2020**, *378*, 20200066. [CrossRef] [PubMed]
2. de Groot, S.R.; Mazur, P. *Non-Equilibrium Thermodynamics*; North-Holland Publishing Company: Amsterdam, The Netherlands, 1962.
3. Hyon, Y.; Kwak, D.Y.; Liu, C. Energetic variational approach in complex fluids: Maximum dissipation principle. *Discret. Contin. Dyn. Syst.* **2017**, *26*, 1291–1304. [CrossRef]
4. Jou, D.; Casas-Vázquez, J.; Lebon, G. *Extended Irreversible Thermodynamics*, 4th ed.; Springer: New York, NY, USA, 2010.
5. Lebon, G.; Jou, D.; Casas-Vázquez, J. *Understanding Non-Equilibrium Thermodynamics: Foundations, Applications*; Springer: London, UK, 2008.
6. Müller, I.; Ruggeri, T. *Rational Extended Thermodynamics*; Springer: New York, NY, USA, 1998.
7. Öttinger, H.C. *Beyond Equilibrium Thermodynamics*; Wiley-Interscience: Hoboken, NJ, USA, 2005.
8. Zhu, Y.; Hong, L.; Yang, Z.; Yong, W.-A. Conservation-dissipation formalism of irreversible thermodynamics. *J. Non-Equilib. Thermodyn.* **2015**, *40*, 67–74. [CrossRef]
9. Yong, W. A. Intrinsic properties of conservation-dissipation formalism of irreversible thermodynamics. *Philos. Trans. R. Soc. A* **2020**, *378*, 20190177. [CrossRef] [PubMed]
10. Pavelka, M.; Klika, V.; Grmela, M. *Multiscale Thermo-Dynamics*; de Gruyter: Berlin, Germany, 2018.
11. Kreiss, H.-O.; Lorenz, J. Initial-boundary value problems and the Navier-Stokes equations. In *Pure and Applied Mathematics*; Academic Press, Inc.: Boston, MA, USA, 1989; Volume 136, p. xii+402.
12. Szücs, M.; Kovács, R.; Simic, S. Open Mathematical Aspects of Continuum Thermodynamics: Hyperbolicity, Boundaries and Nonlinearities. *Symmetry* **2020**, *12*, 1469. [CrossRef]
13. Yong, W.-A. Singular Perturbations of First-Order Hyperbolic Systems. PhD Thesis , Universität Heidelberg, Heidelberg, Germany, 1992.
14. Yong, W.-A. Singular perturbations of first-order hyperbolic systems with stiff source terms. *J. Differ. Equ.* **1999**, *155*, 89–132. [CrossRef]

15. Yang, Z.; Yong, W.-A. Validity of the Chapman-Enskog expansion for a class of hyperbolic relaxation systems. *J. Differ. Equ.* **2015**, *258*, 2745–2766. [CrossRef]
16. Cai, Z.; Fan, Y.; Li, R. Globally hyperbolic regularization of Grad's moment system in one dimensional space. *Commun. Math. Sci.* **2013**, *11*, 547–571. [CrossRef]
17. Benzoni-Gavage, S.; Serre, D. *Multidimensional Hyperbolic Partial Differential Equations: First Order Systems and Applications*; Clarendon Press: Oxford, UK, 2007.
18. Kreiss, H.-O. Initial boundary value problems for hyperbolic systems. *Comm. Pure Appl. Math.* **1970**, *23*, 277–298. [CrossRef]
19. Yong, W.-A. An interesting class of partial differential equations. *J. Math. Phys.* **2008**, *49*, 033503. [CrossRef]
20. Yong, W.-A. Basic aspects of hyperbolic relaxation systems. In *Advances in the Theory of Shock Waves*; Freistühler H., Szepessy, A., Eds.; Progress in Nonlinear Differential Equations and Their Applications; Birkhäuser: Boston, MA, USA, 2001; Volume 47, pp. 259–305.
21. Majda, A.; Osher, S. Initial-boundary value problems for hyperbolic equations with uniformly characteristic boundary. *Comm. Pure Appl. Math.* **1975**, *28*, 607–675. [CrossRef]
22. Gustafsson, B.; Kreiss, H.O.; Oliger, J. *Time-Dependent Problems and Difference Methods*, 2nd ed.; John Wiley & Sons, Inc.: Hoboken, NJ, USA, 2013.
23. Yong, W.-A. Boundary conditions for hyperbolic systems with stiff source terms. *Indiana Univ.Math. J.* **1999**, *48*, 115–137. [CrossRef]
24. Hersh, R. Mixed problems in several variables. *J. Math. Mech.* **1963**, *12*, 317–334.
25. Zhao, W.; Yong, W.-A. Boundary conditions for kinetic theory based models II: A linearized moment system. *Math. Methods Appl. Sci.* 2021. accepted. [CrossRef]
26. Zhou, Y.; Yong, W.-A. Boundary conditions for hyperbolic relaxation systems with characteristic boundaries of type I. *J. Differ. Equ.* **2021**, *281*, 289–332. [CrossRef]
27. Zhou, Y.; Yong, W.-A. Boundary conditions for hyperbolic relaxation systems with characteristic boundaries of type II. submitted.
28. Kato, T. *A Short Introduction to Perturbation Theory for Linear Operators*; Springer: New York, NY, USA, 1982.
29. Făciu, C.; Mihăilescu-Suliciu, M. The energy in one-dimensional rate-type semilinear viscoelasticity. *Int. J. Solids Struct.* **1987**, *23*, 1505–1520. [CrossRef]
30. Zhou, Y.; Yong, W.-A. Construction of boundary conditions for hyperbolic relaxation approximations I: The lin-earized Suliciu model. *Math. Model. Methods Appl. Sci.* **2020**, *30*, 1407–1439. [CrossRef]

MDPI
St. Alban-Anlage 66
4052 Basel
Switzerland
Tel. +41 61 683 77 34
Fax +41 61 302 89 18
www.mdpi.com

Symmetry Editorial Office
E-mail: symmetry@mdpi.com
www.mdpi.com/journal/symmetry